# Chemistry Research and Applications

nova science publishers

www.novapublishers.com

# Chemistry Research and Applications

**The Chemistry of Gallic Acid and Its Role in Health and Disease**
Jeff C. Murdoch (Editor)
2023. ISBN: 979-8-88697-672-4 (Softcover)
2023. ISBN: 979-8-88697-729-5 (eBook)

**A Review of Hydrazine and Its Applications**
Thomas W. Gerdes (Editor)
2023. ISBN: 979-8-88697-671-7 (Hardcover)
2023. ISBN: 979-8-88697-678-6 (eBook)

**Pyrene: Chemistry, Properties and Uses**
**Charles R. Howe** (Editor)
2023. ISBN: 979-8-88697-670-0 (Softcover)
2023. ISBN: 979-8-88697-677-9 (eBook)

**Pyrimidines and their Importance**
Roger G. Ward (Editor)
2023. ISBN: 979-8-88697-656-4 (Softcover)
2023. ISBN: 979-8-88697-663-2 (eBook)

**What to Know about Lanthanum**
Catherine C. Bradley (Editor)
2023. ISBN: 979-8-88697-615-1 (Softcover)
2023. ISBN: 979-8-88697-623-6 (eBook)

**The Future of Biorefineries**
Waldemar Nyström (Editor)
2023. ISBN: 979-8-88697-524-6 (Hardcover)
2023. ISBN: 979-8-88697-528-4 (eBook)

More information about this series can be found at
https://novapublishers.com/product-category/series/chemistry-research-and-applications/

**Barry Schneider**
Editor

# The Chemistry of Pyrimidine Derivatives

nova
science publishers
www.novapublishers.com

**Library of Congress Cataloging-in-Publication Data**

ISBN: 979-8-89113-563-5

Published by Nova Science Publishers, Inc. † New York

# Contents

# Preface

This book contains six chapters on the chemistry of pyrimidine derivatives. Chapter one sums up the pharmacological potential of pyrimidine derivatives and provides an overview of recent studies of these unique compounds as anti-inflammatory agents. Chapter two covers various pyrimidine derivatives with powerful biological and pharmacological applications. The authors in chapter three answer the question of how pyrimidine bases formed on the Earth, i.e., how they evolved. Chapter four begins with an introduction to pyrimidine's structure and then delves into the diverse reactions of pyrimidine derivatives. Chapter five presents the applications of pyrimidine derivatives in the adsorption of heavy metals, which are environmental pollutants. The final chapter provides readers with an assessment of the molecular targets linked to cancer, aiming to clarify their role in disease advancement and the possibility for treatment.

# Chapter 1

# Pyrimidine Derivatives as Anti-Inflammatory Agents

**Halil İbrahim Çetintaş[1]**
**and Burak Tüzün[2,*]**

[1]Advanced Technology Research and Application Center (CUTAM), Sivas Cumhuriyet University, Sivas, Turkey
[2]Plant and Animal Production Department, Technical Sciences Vocational School of Sivas, Sivas Cumhuriyet University, Sivas, Turkey

## Abstract

Although the discovery of a new drug is time-consuming and can cost billions of dollars, it is highly essential for improving patient outcomes with minimum toxicity. Therefore, pyrimidine-based agents have been extensively studied for their potential as anti-inflammatory drugs in addition to their unique chemical and biological properties.

Just like pyridine, pyrimidine is a heterocyclic organic compound that consists of two nitrogen atoms in a six-membered ring. Pyrimidine derivatives have been found to possess anti-inflammatory activities through inhibition of the production of prostaglandins, nitric oxide, nuclear factor kappa-light-chain-enhancer of activated B cells, chemokines, and cytokines, which are important mediators of inflammation.

Due to their mechanism of action and effectiveness, pyrimidine derivatives are promising candidates for the development of new drugs to treat various inflammatory diseases. This chapter sums up the pharmacological potential of pyrimidine derivatives and provides an overview of recent literature studies of these unique compounds as anti-inflammatory agents.

---

* Corresponding Author's Email: theburaktuzun@yahoo.com.

In: The Chemistry of Pyrimidine Derivatives
Editor: Barry Schneider
ISBN: 979-8-89113-563-5

**Keywords:** pyrimidine derivatives, anti-inflammatory activity, medicinal drugs, heterocyclic compounds, inflammation

## 1. Introduction

Heterocyclic compounds are among the molecular structures commonly used in pharmaceutical medicine. In particular, pyrimidines and their derivatives are of great interest due to their wide biological potential. Pyrimidines are organic heterocyclic compounds with a six-membered unsaturated ring structure and usually contain two nitrogen atoms at positions 1 and 3 (Figure 1). These nitrogen atoms are important groups that affect the chemical and biological properties of pyrimidines (Tomma, Khazaal, and Al-Dujaili 2014). It is known that a large number of natural and synthetic pyrimidines exist. The most abundant pyrimidines are uracil, cytosine, and thymine. DNA and RNA bases are the most commonly identified pyrimidine bases, and the pyrimidines have a unique place in chemistry due to their nucleic acid component. They exhibit a range of pharmacological effects, including antibacterial, antiviral, antifungal, antituberculosis, antioxidant, and anti-inflammatory (Rani et al. 2016; Rashid et al. 2021).

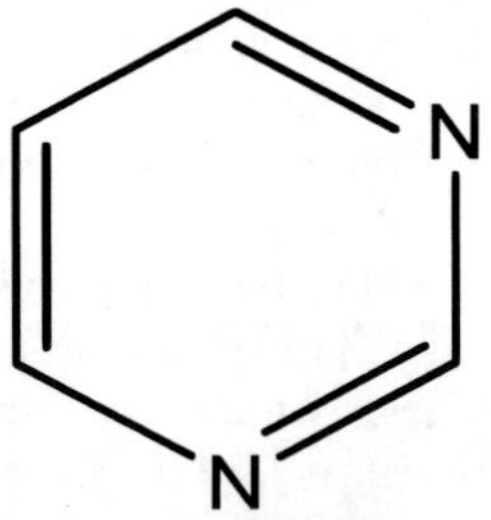
N
N

**Figure 1.** Pyrimidine structure.

Inflammation is a biological response that can be triggered by a variety of factors. These factors can be various, such as damaged cells, pathogens, and toxic compounds. The organs affected by inflammation include the heart, pancreas, liver, kidney, lung, brain, intestinal tract, and reproductive system. Inflammatory responses in these areas can be acute or chronic (Chen et al. 2018). Acute inflammation grows rapidly and soon becomes severe. Although the symptoms may last for a few days, in some cases, it is possible for them to last for weeks. The most common symptoms of acute inflammation include swelling, redness, pain, difficulty moving, and fever. Chronic inflammation, on the other hand, is a long process in which active inflammation, tissue damage, and repair occur simultaneously, lasting several months and sometimes even years. Often, the extent and effects of chronic inflammation vary depending on the source of the injury and the body's effectiveness in healing and controlling the damage. Symptoms of chronic inflammation include body pain, fever, rash, weight gain or loss, fatigue, joint pain, and mouth sores (Rashid et al. 2021).

Today, nonsteroidal anti-inflammatory drugs (NSAIDs), such as aspirin and ibuprofen, are widely used to relieve pain caused by inflammation. These drugs act by inhibiting the activity of an enzyme called cyclooxygenase (COX), which plays a crucial role in the production of biologically active substances known as prostanoids. Prostanoids are involved in a variety of physiological processes, including inflammation. Studies to date have shown that pharmacologically inhibiting COX can provide relief from inflammatory pain. However, the use of NSAIDs causes side effects such as stomach ulcers, gastric corrosion, and kidney toxicity. For this reason, researchers have focused on developing new anti-inflammatory drugs that can be alternatives to NSAIDs with little or no side effects (Day and Graham 2004; Rashid et al. 2019).

Due to their high biological efficacy and anti-inflammatory activity, researchers have focused on pyrimidine derivatives, which can be an effective and safe alternative to NSAIDs, especially in long-term use.

## 2. Pharmacological Potential of Pyrimidine Derivatives

In drug research and development, commercializing a new drug is a long and expensive process, with an average cost of over 1–2 billion dollars. The high cost of these drugs, which require a period ranging from 5 to 15 years from the discovery of the active ingredient to the final stage of human trials, is mostly associated with the development period of new drugs. Also, the high cost is due to the low success rates of candidate molecules, where usually only one out of every 10,000 molecules can be marketed. Therefore, it is extremely important to search for classes of compounds that offer promising pharmacological activities. Because of their potential pharmaceutical properties, pyrimidines have attracted considerable attention among the various classes of compounds that have been widely studied (Sun et al. 2022; Caique Santos Costa et al. 2023).

Numerous natural compounds, including nucleotides, nucleic acids, vitamins, pterins, and antibiotics, contain pyrimidines as structural elements. In all organisms, pyrimidine nucleotides are involved in essential processes in nucleic acids and cell metabolism, including sugar activation for polysaccharide and phospholipid synthesis or glycosylation of proteins and lipids. In addition, they have been demonstrated to be involved in numerous extracellular processes, such as modulating the tone of vascular smooth muscle and functioning as neurotransmitters and neuromodulators in nervous systems (Löffler and Zameitat 2013).

Over the years, heterocyclic compounds have played an important role in the field of medicinal chemistry due to their diverse application potential. It is noteworthy that nitrogen-containing heterocyclic compounds have taken place even in the early studies of chemistry (Ajani et al. 2015). Pyrimidines were already known as degradation products of uric acid even in the very early days of organic chemistry. Alloxan, synthesized in 1818 by Brugnatelli by the oxidation of uric acid with nitric acid, was the first pyrimidine derivative to be isolated (Lagoja 2005). Heterocyclic compound studies constitute almost 50% of all current research in organic chemistry. The basis of this situation is that heterocyclic structures form the basis of different products developed in many fields, such as agriculture, pharmaceuticals, and veterinary medicine. Especially in the field of medicinal chemistry, pyrimidines stand out among the existing heterocyclic compounds due to their broad range of pharmacological activities (Zarenezhad, Farjam, and Iraji 2021), including anti-cancer (Kantankar et al. 2021), anti-viral (Nagalakshmamma et al. 2020), anti-microbial (Dave and Rahatgaonkar 2016), analgesic (Hanna 2012), anti-oxidant (Chkirate et al. 2020), anti-diabetic (Peytam et al. 2021), anti-depressant (Fioravanti et al. 2023), and anti-inflammatory (Wang et al. 2022).

Pyrimidine and its derivatives have made great contributions both to the prevention and treatment of many types of cancer. Most current cancer treatments have many side effects that affect human health and can cause serious health problems (Patil 2023). For this reason, researchers have focused on developing more effective cancer treatments with fewer side effects.

Sivaiah et al. (2023) synthesized a new series of pyrazolo[1,5-*a*]pyrimidine derivatives and investigated them for their anti-cancer potential against HepG2 (liver), MCF-7 (breast), and HCT116 (colorectal) cell lines. The molecular structures were confirmed by NMR, IR, LC/MS, and elemental analysis. Among the total 15 molecules with $IC_{50}$ values ranging from 3.19 to 15.97μM, compounds 1 and 2 (Figure 2) demonstrated the best anti-cancer activity against the MCF-7 cell line with a better safety profile than sorafenib (Sivaiah et al. 2023).

1

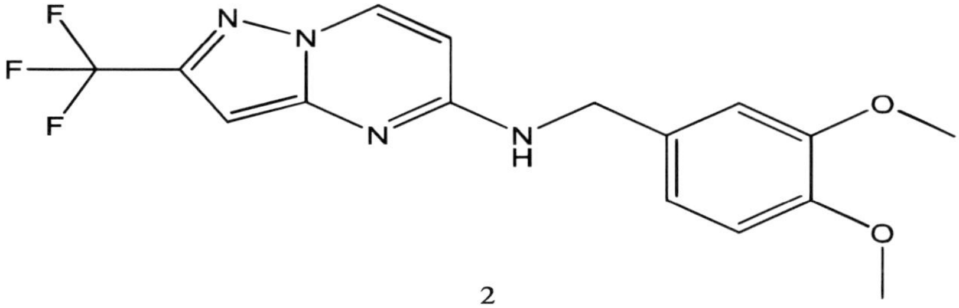

2

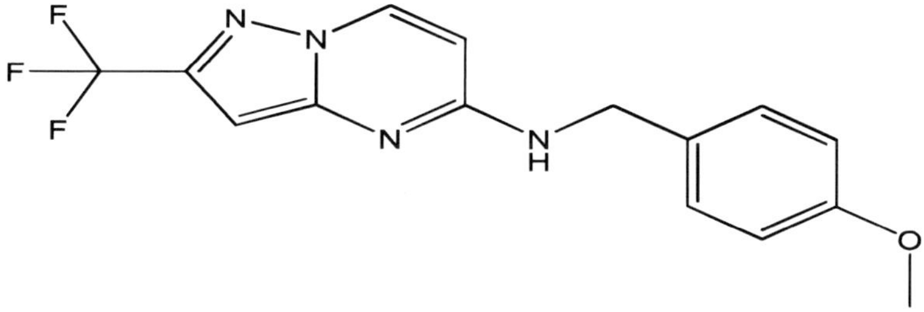

**Figure 2.** Chemical structure of pyrimidine derivatives 1 and 2.

The majority of antiviral nucleoside compounds act as mimics of natural nucleosides to inhibit viral genome replication (Meneghesso et al. 2012). Inhibition also occurs when antiviral agents target viral proteins (Nagalakshmamma et al. 2020). Starting with pyrazolo[3,4-b]pyridin-6-yl)-N,N-dimethylcarbamimidic chloride, Alamshany et al. (2023) produced novel pyrimidine derivatives by using various chemical processes and screened the anti-viral properties of the molecules against SARS-CoV-2. The synthesized compounds were confirmed through MS, IR, NMR, and elemental analysis, as well as molecular docking studies. They reported that the synthesized compounds **3**, **4**, and **5** (Figure 3) displayed the most promising anti-viral efficacy against SARS-CoV-2 and had lower $IC_{50}$ values than lopinavir (Alamshany et al. 2023).

Despite the numerous medicines now on the market, it is of high importance to develop novel anti-microbial therapeutics for the effective treatment of various microbial diseases. In this context, there are many studies showing that pyrimidines are effective antimicrobial agents. Sivagami et al. (2022) synthesized a series of cyanoimino pyrimidine derivatives with different side chains in order to screen in vitro anti-microbial activity against specific bacteria and confirmed the chemical structures using MS, IR, and NMR analyses. They reported that most of the synthesized compounds exhibited antibacterial activity against bacterial and fungal strains. The results revealed that compound **6** (Figure 4) had the strongest impact on *K. pneumoniae*, a gram-negative bacterium, compared to the reference drug Moxifloxacin (Sivagami et al. 2022).

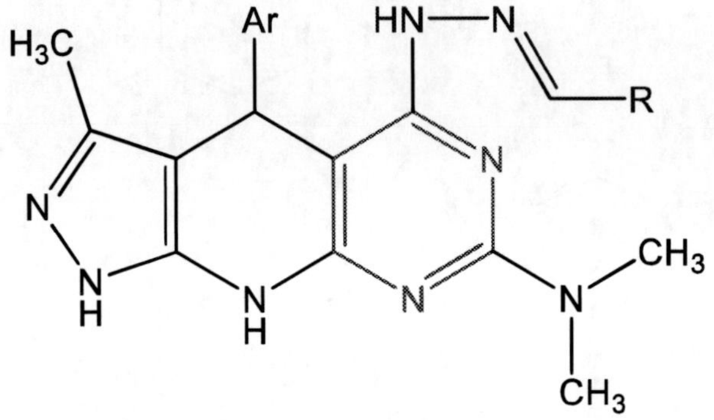
H3C
Ar
HN
N
R
N
N
H
N
H
N
CH3
N
CH3

**3**: $C_6H_4Cl$

**4**: $C_6H_4Br$

5: $C_6H_4Me$

**Figure 3.** Chemical structure of pyrimidine derivatives 3, 4, and 5.

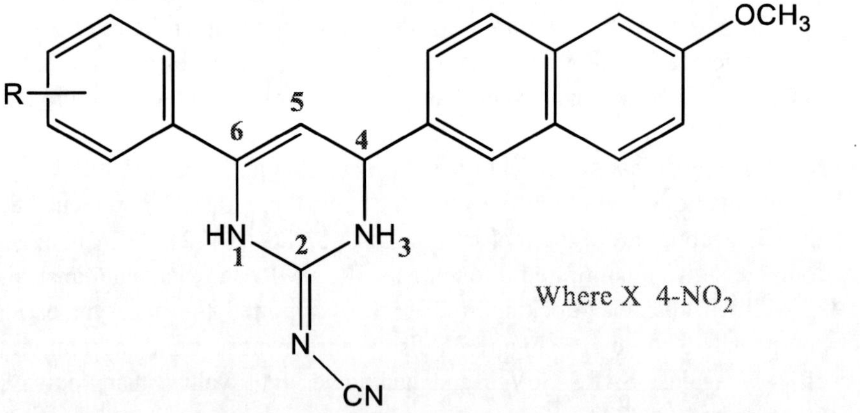
OCH3
R
5
6
4
HN
1
2
NH
3
N
CN

Where X 4-$NO_2$

**Figure 4.** Chemical structure of pyrimidine derivative 6.

Pain is an intense sensation resulting from a noxious sensory stimulus that warns the body of possible tissue and organ damage (Keri et al. 2010). Existing treatments for pain, unfortunately, often have limited efficacy, and some drugs cause side effects that make their clinical use problematic and prevent their long-term use. Therefore, it is necessary to develop safer and more effective methods for the treatment of pain. A new series of 2,5-substituted 4-(trifluoromethyl)-spirochromeno[4,3-d]pyrimidines were designed and synthesized by Bonacorso et al. (2017). Chemical structures of the compounds were characterized by NMR, GC/MS, HRMS and elemental analysis, and evaluated their analgesic potential in mouse pain model. The findings demonstrated that compound 7 (Figure 5) had a high potential for novel analgesic medications in the treatment of pathological pain in arthritis (Bonacorso et al. 2017).

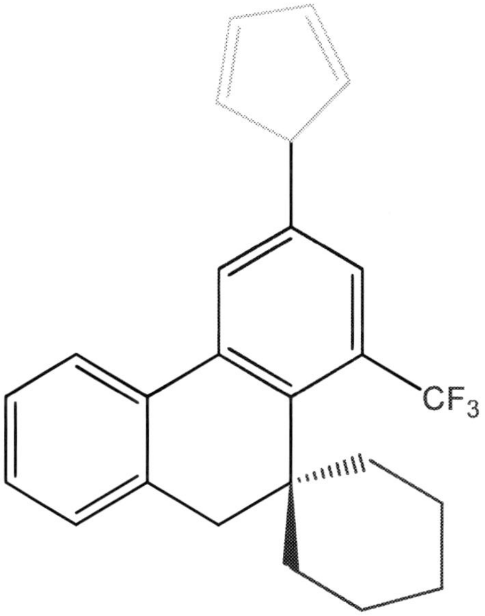
$CF_3$

**Figure 5.** Chemical structure of pyrimidine derivative 7.

In medicinal chemistry, anti-oxidants are compounds that can help collect or scavenge radicals to preserve biological tissues from extraneous harm (Wang et al. 2023). A novel series of 6-oxo-1,6-dihydropyrimidin-5-carboxamides were synthesized in high yield and evaluated for their anti-oxidant potential by Rambabu et al. (2021). The compounds were confirmed by IR, NMR, and HRMS analyses, and then their in vitro anti-oxidant activities were investigated through the microdilution technique. The results indicated that the compounds **8**, **9**, and **10** (Figure 6) showed very high anti-oxidant activity against ascorbic acid (Rambabu et al. 2021).

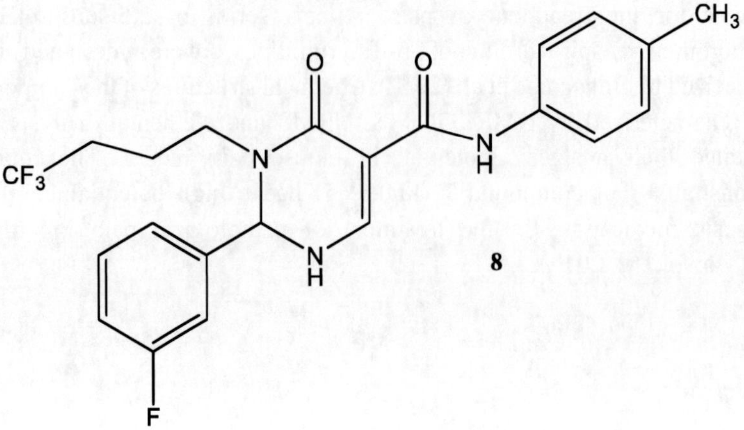
CH3
O
O
N
H
CF3
N
N
H
F

**8**

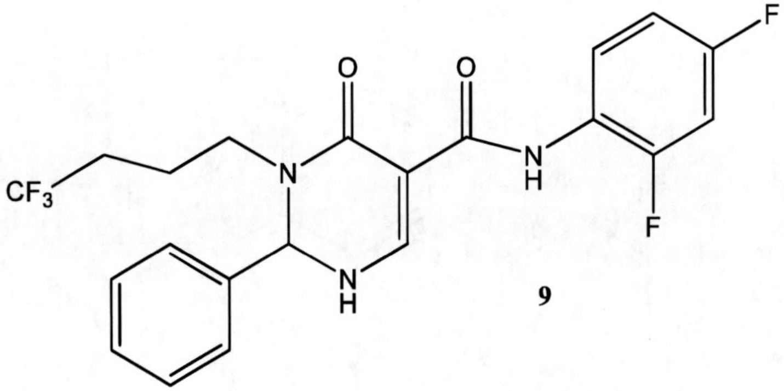
F
O
O
N
H
CF3
N
F
N
H

**9**

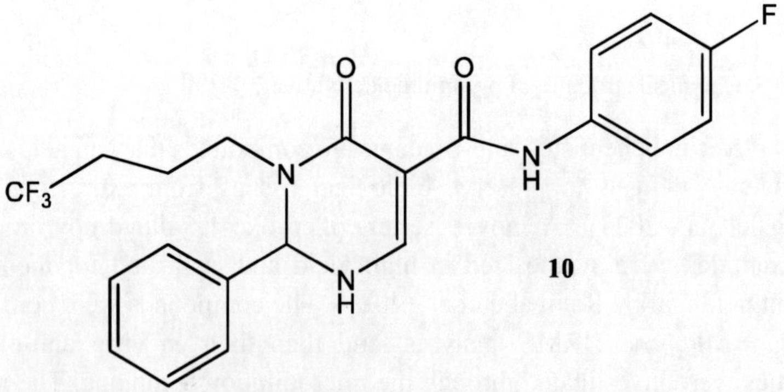
F
O
O
N
H
CF3
N
N
H

**10**

**Figure 6.** Chemical structure of pyrimidine derivatives 8, 9, and 10.

In the 21st century, diabetes mellitus is one of the most common metabolic disorder characterized by chronic hyperglycemia. This disease, which causes serious health problems, including cardiovascular diseases, kidney damage and neuropathy, has high mortality rates (Mushtaq et al. 2023). Therefore, the search for new and more potent inhibitors with fewer side effects and lower costs is of great importance. Recently, Amin et al. (2023) reported a new series of pyrimidine-thiazolidinedione derivatives and characterized their chemical structures by IR, NMR, and MS techniques. All the compounds were investigated for in vivo anti-diabetic evaluation in streptozotocin-induced diabetic rats, and the results demonstrated that compounds **11** and **12** (Figure 7) dramatically reduced glucose levels in the blood after four weeks of administration. The results showed that the pyrimidine-based thiazolidinedione derivatives have potent anti-diabetic activity with few side effects (Amin et al. 2023).

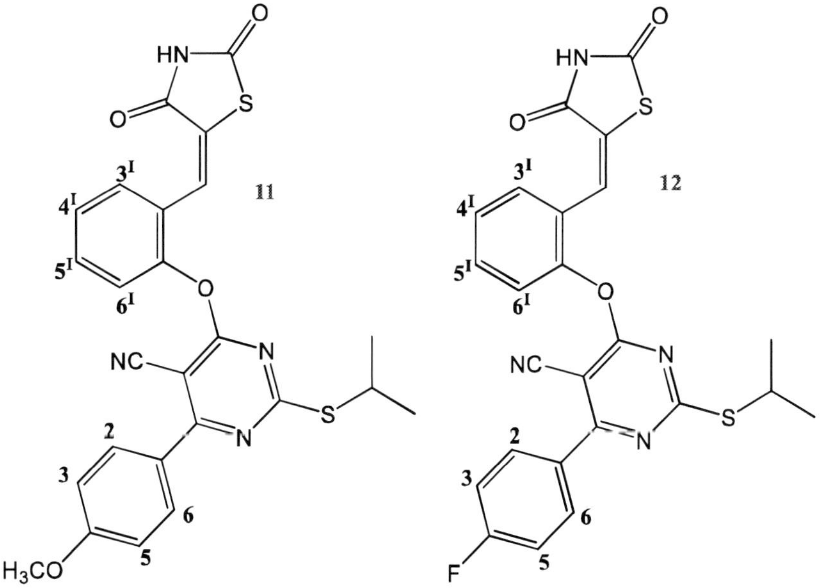
HN
O
S
O
3I
4I
5I
6I
11
O
NC
N
S
N
2
3
6
5
H3CO
HN
O
S
O
3I
4I
5I
6I
12
O
NC
N
S
N
2
3
6
5
F

**Figure 7.** Chemical structure of pyrimidine derivatives 11 and 12.

Depression is a prevalent mental disorder that involves an intricate interplay between social and psychological factors, impacting approximately 264 million individuals globally, as reported by the World Health Organization (WHO). Numerous scientific groups have endeavored to develop potential drugs that are able to interact effectively with receptors. The most frequently employed approach in drug design involves integrating various structural elements widely acknowledged as having anti-depressant features into the synthesized compounds, aiming to augment their central nervous system effects (Singh et al. 2021). In this context, Fioravanti et al. (2023) synthesized a series of novel pyrimidine thioether derivatives in order to discover alternative analogues of the antihypoxic medication Isothiobarbamine and investigated them both in silico and in vivo. The in vivo experiments resulted in the discovery of several compounds that exhibited antidepressant/anxiolytic, performance-enhancing, and nootropic characteristics. Further testing of compounds **13** and **14** (Figure 8) on 400 socially depressed albino mice demonstrated increased motor and exploratory activity compared to control groups. The treated mice also displayed a higher frequency of social interaction and better results in a sucrose preference test, indicating an improved psychoemotional state. Additionally, compounds **13** and **14** showed minimal acute toxicity, which was lower than that of Fluoxetine hydrochloride (Fioravanti et al. 2023).

13

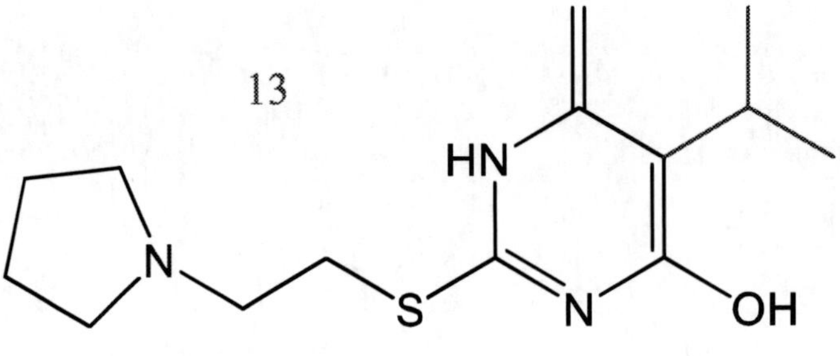
HN
N
S
N
OH

14

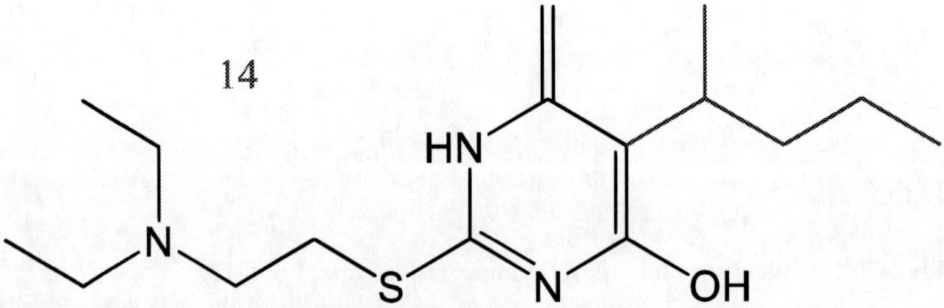
HN
N
S
N
OH

**Figure 8.** Chemical structure of pyrimidine derivatives 13 and 14.

## 3. Recent Advances in the Anti-Inflammatory Properties of Pyrimidines

Inflammation is a protective reaction that occurs in response to harmful stimuli or conditions, such as infection and tissue damage (Medzhitov 2008). Its purpose is mainly to remove harmful stimuli and initiate the healing process. The characteristic signs of inflammation can be attributed to increased blood flow, heightened cellular metabolism, vasodilation, the release of certain substances, the leakage of fluids, and the influx of cells. At the tissue level, inflammation is identified by redness, swelling, heat, pain, and impaired tissue function. Inflammation can be either acute or chronic. The innate immune system plays a pivotal role in mediating the initial response (Nielsen, Andersen, and Girardin 2007; Chen et al. 2018). Typically, in acute inflammatory responses, cellular and molecular processes work effectively to reduce possible injury or infection. This process helps restore tissue balance and resolve the acute inflammation. However, if acute inflammation is uncontrolled, it can progress to a chronic state and contribute to the development of various chronic inflammatory diseases (Chen et al. 2018). To alleviate pain and reduce fever associated with inflammatory conditions, nonsteroidal anti-inflammatory drugs (NSAIDs) such as ibuprofen, aspirin, or naproxen can be employed, as well as herbal supplements like curcumin, capsaicin, and Boswellia serrata, and corticosteroids like prednisone (Rashid et al. 2021).

The inflammatory response comprises the coordinated activation of signaling pathways that regulate the levels of inflammatory mediators in both tissue cells and inflammatory cells recruited from the blood. Inflammation serves as a common underlying mechanism for many chronic disorders, including cardiovascular disorders, bowel diseases, diabetes, arthritis, and cancer. Although the specific processes of the inflammatory response can vary depending on the nature and location of the initial stimulus, they all share a common mechanism, which can be summarized in the following sequence: recognition of harmful stimuli by cell surface pattern receptors; activation of inflammatory pathways; release of inflammatory markers; and recruitment of inflammatory cells (Chen et al. 2018).

Non-steroidal anti-inflammatory drugs (NSAIDs) are one of the most commonly used drug families today due to their anti-inflammatory, analgesic, and antipyretic effects. NSAIDs encompass various drug groups with heterogeneous chemical structures. Their mechanism of action involves inhibiting cyclooxygenase (COX) in the arachidonic acid metabolism, thereby reducing prostaglandin ($PGE_2$) synthesis (Khoo et al. 2023). COX exists in two main isoforms: The first isoform, COX-1, plays a role in maintaining homeostasis, protecting the gastric mucosa, regulating platelet aggregation, and modulating renal microcirculation. The second isoform, COX-2, is found in certain tissues and can be stimulated by various factors, particularly proinflammatory cytokines. Therefore, it is believed that COX-1 plays a much more significant role than COX-2 in the early stages of inflammation (Hijos-Mallada, Sostres, and Gomollón 2022).

NSAIDs, with their effectiveness in treating pain and other inflammatory conditions, are among the most widely consumed drugs worldwide (Tay et al. 2022). However, the toxic nature of NSAIDs limits their use in the treatment of inflammatory diseases (Panchal and Prince Sabina 2023). Hence, there is a pressing need in the field of pharmacological research to discover novel and cost-effective anti-inflammatory agents that have minimal adverse effects.

Due to the remarkable pharmacological activity of pyrimidine derivatives, significant research efforts have been focused on their anti-inflammatory potential. Pyrimidines demonstrate their anti-inflammatory activities by inhibiting key inflammatory mediators such as $PGE_2$, nitric oxide (NO), nuclear factor kappa-light-chain-enhancer of activated B cells (NF-κB), chemokines, and cytokines (Rashid et al. 2021). Recently, Abdelkhalek et al. (2023) synthesized a novel series of thieno[2,3-d]pyrimidine derivatives (Figure 9) and evaluated their anti-inflammatory properties against both 15-LOX and COX-2. The team characterized the final compounds by IR, NMR, GC/MS, UPLC–MS/MS and reported that compound **15** exhibited the highest 15-LOX inhibitory activity and 5 times higher COX-2 selectivity than diclofenac, showed a greater potency in reducing NO ($IC_{50}$ = 7.77 μM) than both celecoxib ($IC_{50}$ = 22.89 μM) and diclofenac ($IC_{50}$ = 25.34). Also, the compounds **16** (SI = 137.37) and **17** (SI = 132.26) exhibited very high COX-2 selectivity (Abdelkhalek et al. 2023).

Cai et al. (2023) designed and synthesized a new series of pyrimidine-5-carboxamide derivatives through a scaffold hopping strategy and investigated their anti-inflammatory drug-like features. The compounds were purified by TLC, and chemical structures were confirmed by NMR and Q-TOF LC/MS. Among all the molecules, compound **18** (Figure 10) was reported as the most promising anti-inflammatory drug candidate with favorable activity and selectivity towards SIK1/2, excellent metabolic stability in the human liver microsome, enhanced in vivo exposure, a suitable rate of plasma protein binding, and potent anti-inflammatory activity in a DSS-induced colitis model.

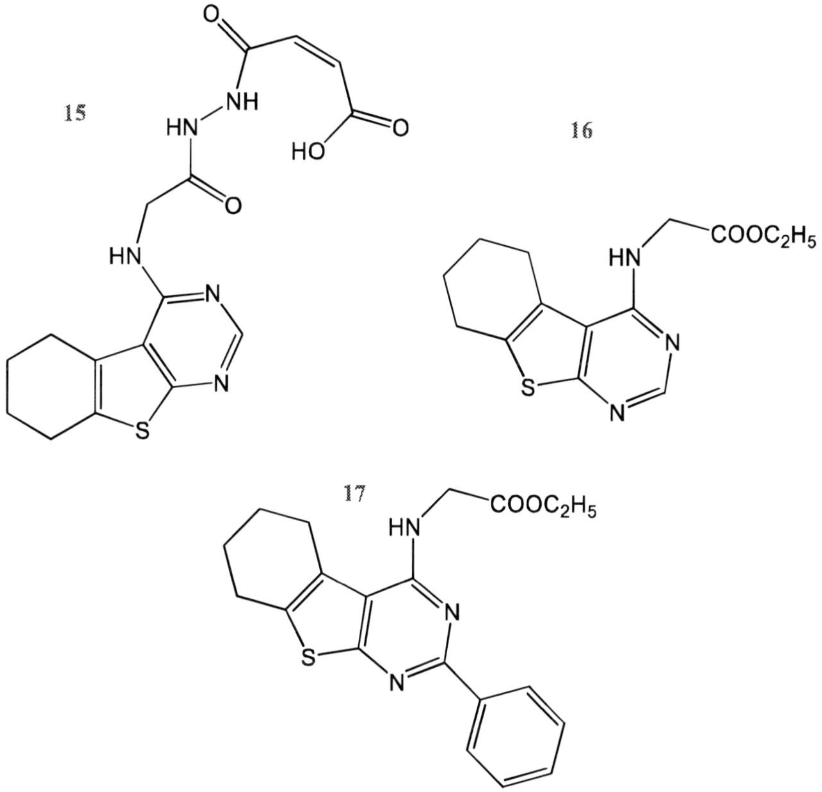
15
O
NH
HN
HO
O
O
HN
N
N
S
16
COOC2H5
HN
N
S
N
17
COOC2H5
HN
N
S
N

**Figure 9.** Chemical structure of pyrimidine derivatives 15, 16, and 17.

In another study, the pyrimidine derivatives were synthesized by combining an appropriate α,β-unsaturated ketone with 4-amino-6-hydroxy-2-mercaptopyrimidine monohydrate in glacial acetic acid by Myriagkou et al. (2023). The compounds were characterized through IR, NMR, LC/MS, and elemental analysis and evaluated for their in vitro anti-inflammatory potential. Among all the tested compounds, pyrimidine derivatives **19** and **20** (Figure 11) exhibited the most potent lipoxygenase inhibition activity with $IC_{50}$ values of 42 and 47.5μM, respectively (Myriagkou et al. 2023).

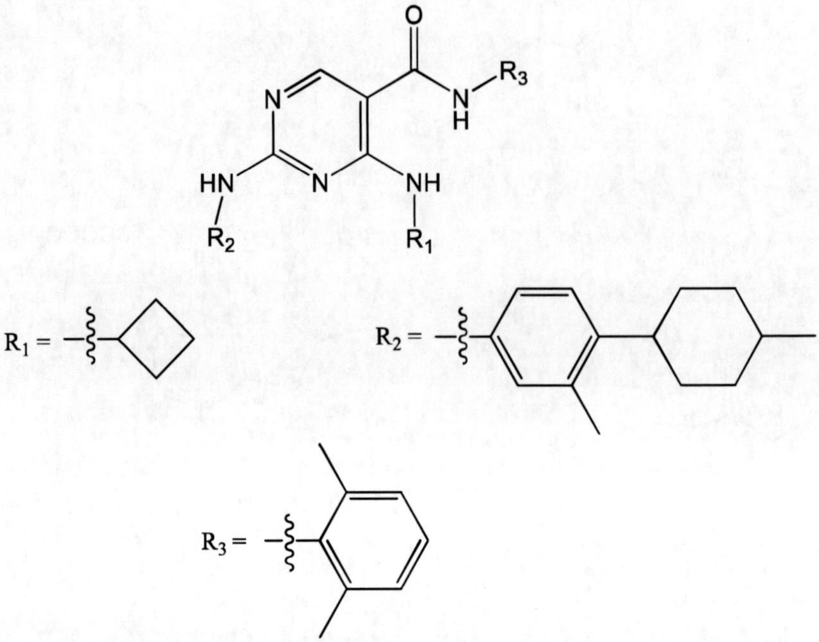
O
N
N
H
R3
HN
N
NH
R2
R1
R1 =
R2 =
R3 =

**Figure 10.** Chemical structure of pyrimidine derivative 18.

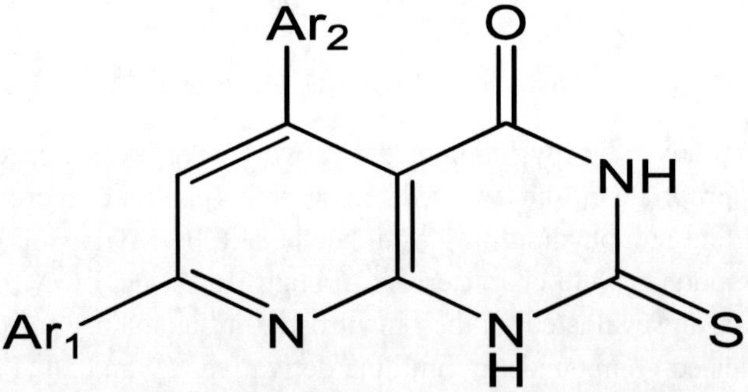
Ar2
O
NH
Ar1
N
N
H
S

**19** $Ar_1 = C_6H_5$ $Ar_2 = C_6H_5$

**20** $Ar_1 = 4\text{-}ClC_6H_4$ $Ar_2 = 4\text{-}ClC_6H_4$

**Figure 11.** Chemical structure of pyrimidine derivatives 19 and 20.

Ayman et al. (2023) designed and synthesized novel pyrazole, imidazo [1,2-b]pyrazole, and pyrazolo [1,5-a]pyrimidine derivatives and evaluated their in vitro anti-inflammatory activity as COX-2 inhibitors. All the chemical structures were confirmed by IR, NMR, GC/MS, and elemental analysis. Two tested derivatives, **21** ($IC_{50}$=5.68 ± 0.08 μM) and **22** ($IC_{50}$=3.37 ± 0.07 μM) (Figure 12), showed high COX-2 inhibition activity compared with celecoxib and meloxicam. In addition, it was confirmed with molecular docking simulation and ADMET predictions that the most potent derivatives exhibited strong binding affinity, possessed suitable drug-like properties, and demonstrated low toxicity profiles (Ayman et al. 2023).

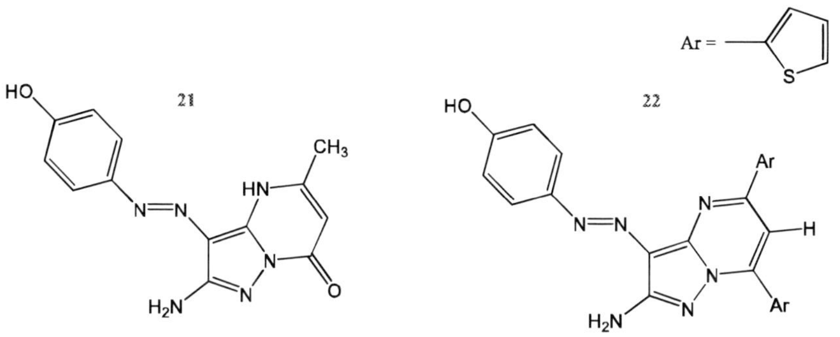
21
22
HO
N=N
HN
CH3
N
O
H2N
Ar =
S
Ar
H

**Figure 12.** Chemical structure of pyrimidine derivatives 21 and 22.

Spasov et al. (2023) identified amino derivatives of 4,6- and 5,7-diaryl substituted pyrimidines and [1,2,4]triazolo[1,5-*a*]pyrimidines as potential NO and interleukin 6 (IL-6) inhibitors and screened their anti-inflammatory efficacy on isolated primary murine macrophages after LPS stimulation. They reported that while 7 out of 14 compounds exhibited NO and IL-6 inhibition activity, the inhibition efficacy of compound **23 (Figure 13)**, the most promising, was validated in an animal model of acute lung injury as evidenced by biochemical, cytological, and morphological markers.

Very recently, a research group found that the pyrimidine derivative **24** (4-(3-chlorophenyl)-1,7-diethylpyrido(2,3-*d*)-pyrimidin-2(1*H*)-one) (Figure 14) effectively reduced lipid accumulation in adipose tissue without causing citotoxicity. This finding was significant since lipid accumulation in adipose tissue is often associated with inflammation. The results of the study demonstrated that compound 24 is a potential novel compound for controlling lipid accumulation and adipocyte formation in obesity, and it also indirectly exhibits anti-inflammatory properties (Kim, Kim, and Um 2023).

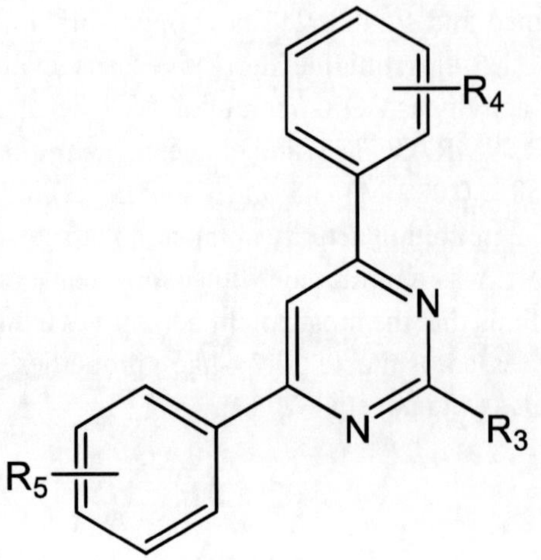
R4
N
N
R3
R5

$R_3 = CH_3$

$R_4 = 4\text{-}NH_2$

$R_5 = 4\text{-}NH_2$

**Figure 13.** Chemical structure of pyrimidine derivative 23.

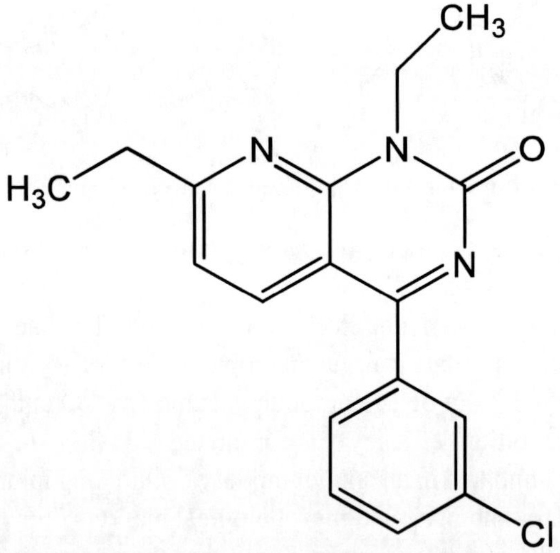
CH3
H3C
N
N
O
N
Cl

**Figure 14.** Chemical structure of pyrimidine derivative 24.

The current medical strategies to combat the COVID-19 pandemic generally involve direct targeting of the SARS-CoV-2 virus through the development of specific drugs and safe vaccines. However, an indirect approach focuses on utilizing anti-inflammatory drugs to control the cytokine storm, which is responsible for severe health complications caused by the virus. In this context, Sayed et al. (2022) synthesized a novel series of fused pyrrolopyrimidine derivatives as potent anti-inflammatory agents with antioxidant properties and confirmed their structures through IR, NMR, GC/MS, and elemental analylsis. The final compounds were evaluated for their in vitro anti-inflammatory potential using RAW264.7 cells. The findings revealed that the molecular docking and simulation results aligned with the biological data, confirming that compounds **25**, **26**, and **27** (Figure 15) possess promising anti-inflammatory activity as well as antioxidant properties (Sayed et al. 2022).

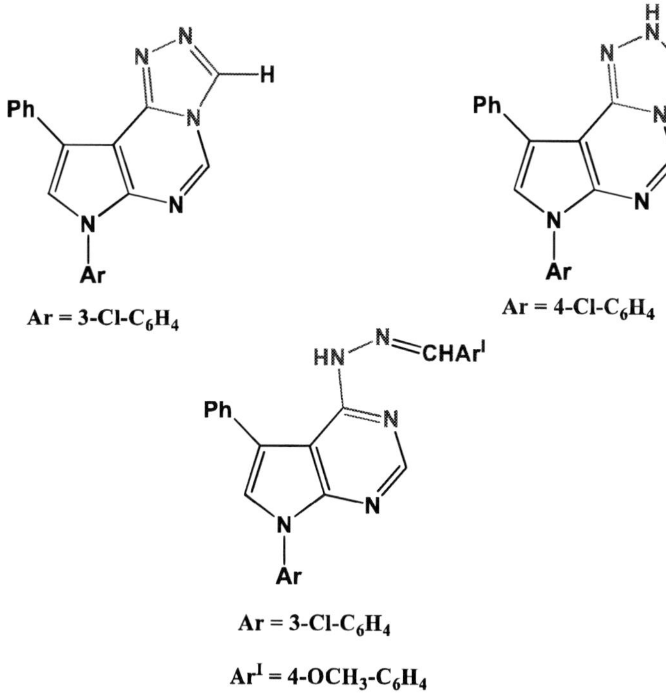
Ar = 3-Cl-C6H4
Ar = 4-Cl-C6H4
Ar = 3-Cl-C6H4
ArI = 4-OCH3-C6H4

**Figure 15.** Chemical structure of pyrimidine derivatives 25, 26, and 27.

Morsy et al. (2022) designed and synthesized a new series of thienopyrimidinone glycoside derivatives and evaluated their anti-inflammatory activity against LOX, COX-1, and COX-2. The final compounds were characterized by IR, NMR, and elemental analysis. Among fourteen in vitro tested compounds, **28**, **29**, and **30** (Figure 16) exhibited potent anti-inflammatory activity compared with ipubrofen and indomethacin standards. Also, the biological properties of these three compounds were confirmed through molecular docking studies, and the findings indicated that these pyrimidine derivatives had dual activity as anti-inflammatory and antimicrobial agents (Morsy et al. 2022).

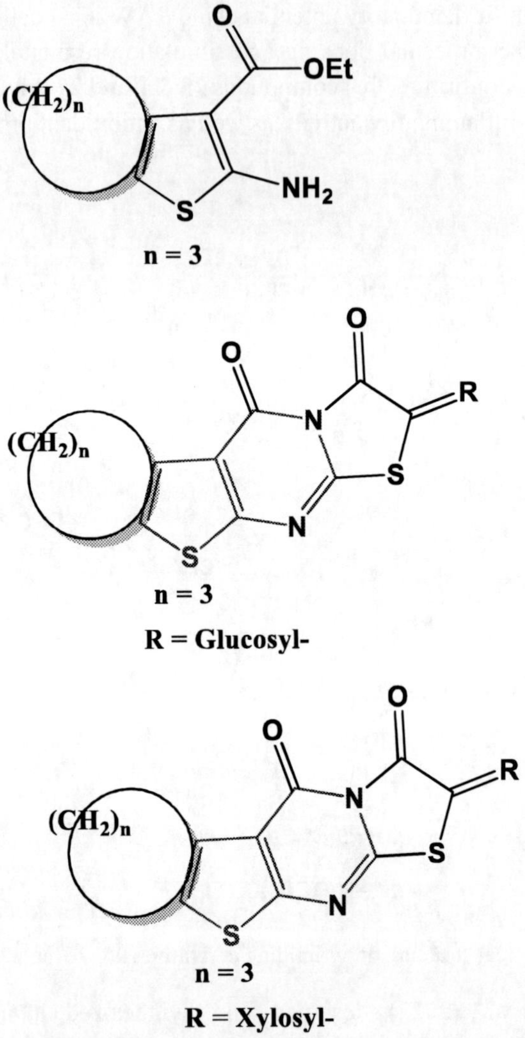
O
OEt
(CH2)n
NH2
S
n = 3
O
O
R
N
(CH2)n
S
N
S
n = 3
R = Glucosyl-
O
O
R
N
(CH2)n
S
N
S
n = 3
R = Xylosyl-

**Figure 16.** Chemical structure of pyrimidine derivatives 28, 29, and 30.

Manzoor et al. (2023) synthesized a new series of triazole-pyrimidine-based compounds and analyzed them using NMR, LC-HRMS, and XRD in order to assess their potential neuroprotective and anti-inflammatory activity through various methods, including cell viability testing (MTT assay), enzyme-linked immunosorbent assay (ELISA), quantitative reverse transcription-polymerase chain reaction (qRT-PCR), western blotting, and molecular docking. Among the fourteen tested compounds, **31–39** (Figure 17) exhibited potent anti-inflammatory activity through inhibition of NO and TNF-α production in LPS-stimulated human microglia cells. The findings showed that triazole-pyrimidine hybrid compounds can be used as potent neuroprotective and anti-inflammatory agents (Manzoor et al. 2023).

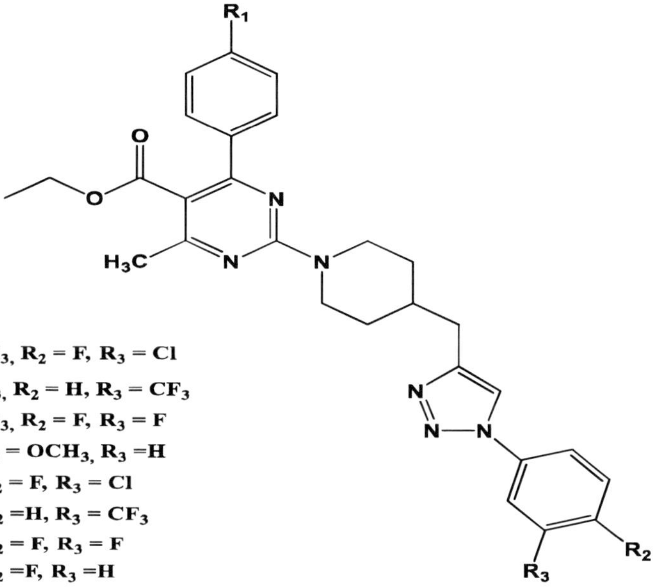

**31 : $R_1 = OCH_3$, $R_2 = F$, $R_3 = Cl$**
**32: $R_1 = OCH_3$, $R_2 = H$, $R_3 = CF_3$**
**33 : $R_1 = OCH_3$, $R_2 = F$, $R_3 = F$**
**34 : $R_1 = H$, $R_2 = OCH_3$, $R_3 = H$**
**35 : $R_1 = H$, $R_2 = F$, $R_3 = Cl$**
**36 : $R_1 = H$, $R_2 = H$, $R_3 = CF_3$**
**37 : $R_1 = H$, $R_2 = F$, $R_3 = F$**
**38 : $R_1 = H$, $R_2 = F$, $R_3 = H$**

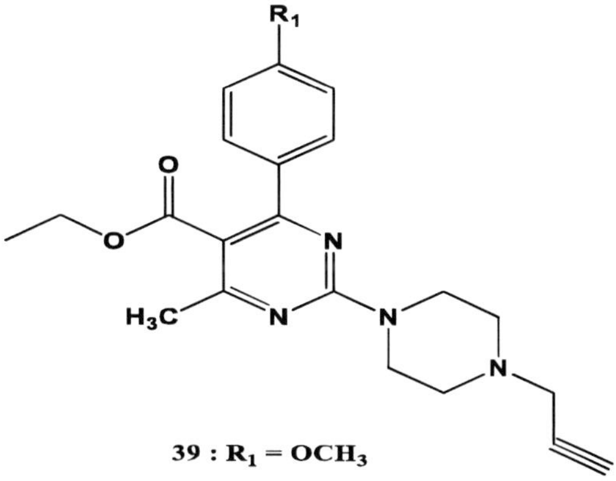

**39 : $R_1 = OCH_3$**

**Figure 17.** Chemical structure of pyrimidine derivatives 31–39.

Similarly, in another study, Wang et al. (2022) designed and synthesized a new series of indole and indazole-piperazine pyrimidine derivatives in order to evaluate their both neuroprotective and anti-inflammatory activities for the treatment of ischemic stroke, and the chemical structures of the compounds were confirmed through NMR and HRMS analyses. The most appealing cytoprotective activity was demonstrated by compound **40** (Figure 18), which also significantly reduced the release of TNF-α and IL-1β in the mouse neuroinflammation model and LPS-induced BV2 cells. The results showed that compound **40** is an efficient inflammatory inhibitor against COX-2 and 5-LOX with $IC_{50}$ values of 92.54 nM and 41.86 nM, respectively, as well as its potent neuroprotective efficacy (Wang et al. 2022).

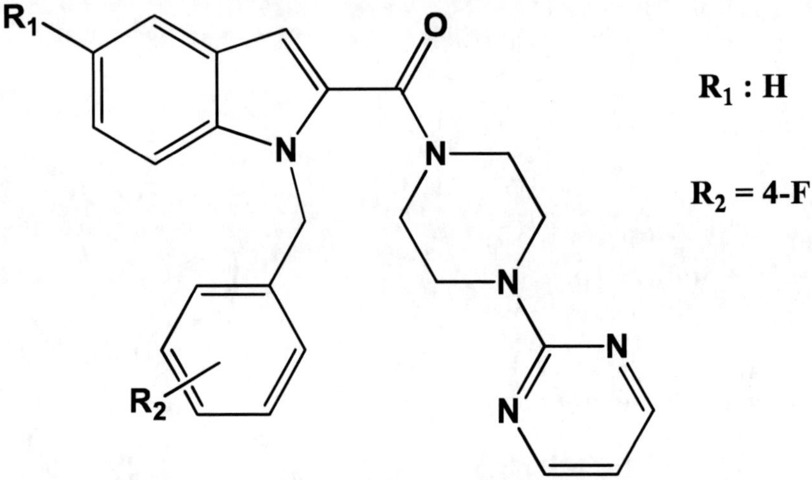
R1 : H
R2 = 4-F

**Figure 18.** Chemical structure of pyrimidine derivative 40.

Khazimullina et al. (2022) employed the molecular docking approach in order to simulate the formation of complexes between seventeen uracil derivatives, which included cyclic and acyclic sulfur- and oxygen-based substituents in the pyrimidine ring with the active sites of cyclooxygenase isoforms (COX). According to the simulation results, compounds **41** and **42** (Figure 19) were identified as having the potential to be effective inhibitors of COX isoforms that are induced during inflammatory processes in the body, particularly with a higher selectivity towards COX-2. When these compounds were screened for their in vivo anti-inflammatory activities in various inflammation models, it was observed that they exhibited significant anti-inflammatory activities, demonstrating efficacy comparable to that of ortofen (Khazimullina et al. 2022).

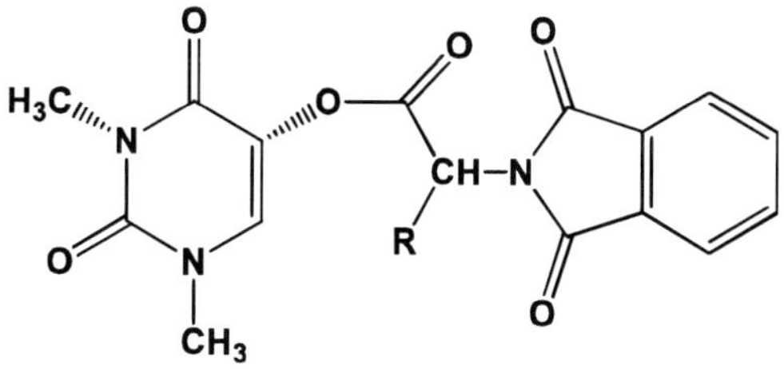

O
H3C
N
O
O
O
CH–N
R
O
N
CH3
O


**41:** R = CH3 **(98%)**
**42 :** R = $CH_2CH_2SCH_3$ **(96%)**

**Figure 19.** Chemical structure of pyrimidine derivatives 41 and 42.

In a recent study, a total of 10 novel (2-hydroxyphenyl)-5-phenyl-6-(pyrrolidine-1-carbonyl)-1H-pyrano[2,3-d]pyrimidine-2,4(3H,5H)-dione derivatives were synthesized by Veeranna et al. (2022) and characterized using IR, NMR, and HRMS techniques. All the derivatives were tested for their in vitro anti-inflammatory properties, and compound **43** (Figure 20) was reported to be the most potent anti-inflammatory molecule with excellent efficacy ($IC_{50}$= 24.500 µµg/mL) compared with standard diclofenac sodium ($IC_{50}$ = 34.24 µg/mL). Also, molecular docking studies were conducted using a COX-2 receptor, and it was indicated that compound **43** holds promise as a potential anti-inflammatory agent in the future (Veeranna et al. 2022).

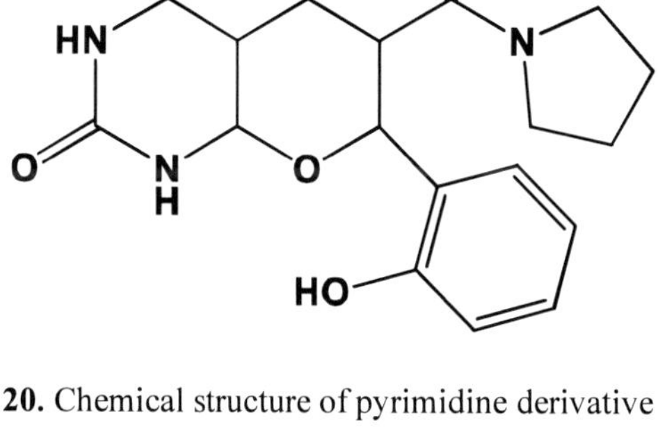

R
O
O
HN
N
O
N
H
O
HO


**R = 4- OCH$_3$**

**Figure 20.** Chemical structure of pyrimidine derivative 43.

Abdelgawad et al. (2022) designed and synthesized a new series of pyrazolo[1,5-*a*] pyrimidine derivatives and evaluated their in vivo and in vitro anti-inflammatory properties. The research group characterized the chemical structures using IR, NMR, EI/MS, and elemental analysis. Among all the synthesized compounds, compound **44** exhibited the highest activity against IL-6 and TNF-α, with percentage inhibitions of 80% and 89%, respectively. Also, compound **45** demonstrated the strongest inhibition of COX-2, with an $IC_{50}$ value of 1.11 μM, while compound **44** displayed the highest selectivity towards COX-2 with a selectivity index (SI) of 8.97. Furthermore, molecular docking simulation studies inside COX-2 and 15-LOX active sites confirmed the anti-inflammatory efficacy of the inhibitory compounds (Figure 21) (Abdelgawad et al. 2022).

**44**

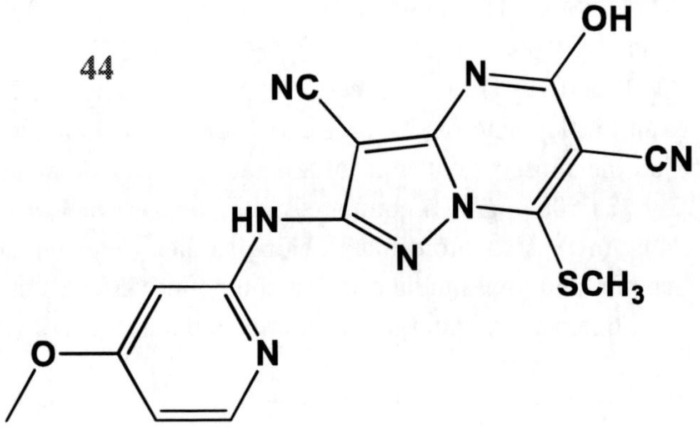
OH
NC
N
CN
HN
N
N
SCH3
O
N

**45**

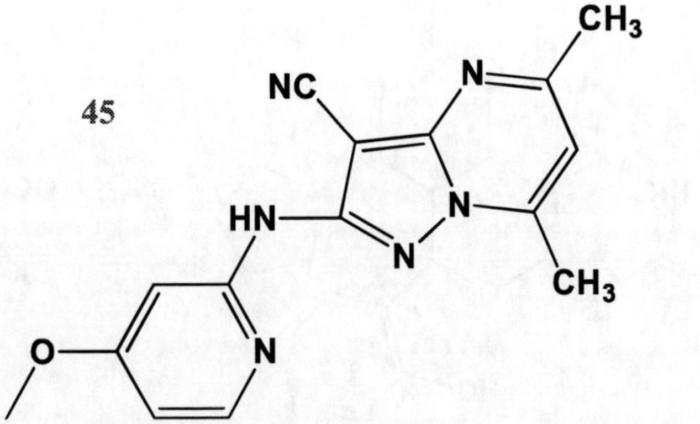
CH3
NC
N
HN
N
N
CH3
O
N

**Figure 21.** Chemical structure of pyrimidine derivatives 44 and 45.

## Conclusion

Inflammation is a part of the body's defense mechanism, and it is activated to eliminate harmful substances, repair damaged tissues, and promote healing. It can be acute or chronic, and NSAIDs are widely used to relieve pain caused by inflammation. However, the utilization of NSAIDs causes various adverse effects. Therefore, researchers have directed their efforts towards the development of novel anti-inflammatory drugs that can serve as alternatives to NSAIDs with little or no side effects.

Heterocyclic compounds, especially pyrimidines and their derivatives, are widely employed in pharmaceutical medicine. These compounds have attracted considerable attention due to their potent biological activity. In this chapter, firstly, a general overview and characteristics of pyrimidines were provided. Then, the pharmacological uses of pyrimidines and their derivatives were briefly summarized. Finally, the recent studies on the use of pyrimidines as anti-inflammatory agents were reviewed in detail.

Pyrimidine derivatives hold promise as prospective candidates for the development of new anti-inflammatory agents. The ongoing intensive research on the use of pyrimidine derivatives as anti-inflammatory agents plays a vital role in advancing effective treatments for various inflammatory conditions. It is now well recognized that these investigations have significant potential to enhance the quality of life for patients by developing effective treatments for diverse inflammatory conditions.

## References


Abdelgawad, Mohamed A., Elkanzi, Nadia A., Musa, Arafa, Ghoneim, Mohammed M., Ahmad, Waqas, Elmowafy, Mohammed, Abdelhaleem Ali, Ahmed M. Abdelazeem, Ahmed H., Bukhari, Syed N. A., El-Sherbiny, Mohamed, Abourehab, Mohammed A. S., Bakr, Rania B. "Optimization of pyrazolo[1,5-a]pyrimidine based compounds with pyridine scaffold: Synthesis, biological evaluation and molecular modeling study." *Arabian Journal of Chemistry 15*, no. 8 (2022): 104015. Accessed July 11, 2023. https://doi.org/10.1016/j.arabjc.2022.104015.

Abdelkhalek, Ahmed S., Kothayer, Hend, Rezq, Samar, Orabi, Khaled Y., Romero, Damian G., and Osama I. El-Sabbagh. "Synthesis of new multitarget-directed ligands containing thienopyrimidine nucleus for inhibition of 15-lipoxygenase, cyclooxygenases, and pro-inflammatory cytokines." *European Journal of Medicinal Chemistry 256*, (2023): 115443. Accessed July 10, 2023. https://doi.org/10.1016/j.ejmech.2023.115443.

Ajani, Olayinka O., Jessica T. Isaac, Taiwo F. Owoeye, and Anuoluwa A. Akinsiku. 2015. "Exploration of the Chemistry and Biological Properties of Pyrimidine as a Privilege Pharmacophore in Therapeutics." *International Journal of Biological Chemistry* 9 (4): 148–77. https://doi.org/10.3923/ijbc.2015.148.177.

Alamshany, Zahra M., Khattab, Reham R., Hassan, Nasser A., A., Ahmed, Tantawy, Mohamed A., Mostafa, Ahmed, and Allam A. Hassan. "Synthesis and Molecular Docking Study of Novel Pyrimidine Derivatives against COVID-19." *Molecules 28*, no. 2 (2023): 739. Accessed July 9, 2023. https://doi.org/10.3390/molecules 28020739.

Amin, Shaista, Sheikh, Khursheed A., Iqubal, Ashif, Ahmed Khan, Mohammad, Shaquiquzzaman, M., Tasneem, Sharba, Khanna, Suruchi, Najmi, A. K., Akhter, Mymoona, Haque, Anzarul, Anwer, Tarique, Alam, M. Mumtaz. "Synthesis, in-Silico studies and biological evaluation of pyrimidine based thiazolidinedione derivatives as potential anti-diabetic agent." *Bioorganic Chemistry 134*, (2023): 106449. Accessed July 10, 2023. https://doi.org/10.1016/j.bioorg.2023.106449.

Ayman, Radwa, Abusaif, Moustafa S., Radwan, A. M., Elmetwally, Amira M., and Ahmed Ragab. "Development of novel pyrazole, imidazo[1,2-b]pyrazole, and pyrazolo[1,5-a]pyrimidine derivatives as a new class of COX-2 inhibitors with immunomodulatory potential." *European Journal of Medicinal Chemistry 249*, (2023): 115138. Accessed July 11, 2023. https://doi.org/10.1016/j.ejmech.2023.115138.

Bonacorso, Helio G., Rosa, Wilian C., Oliveira, Sara M., Brusco, Indiara, Brum, Evelyne S., Rodrigues, Melissa B., Frizzo, Clarissa P., and Nilo Zanatta. "Synthesis of novel trifluoromethyl-substituted spiro-[chromeno[4,3-d]pyrimidine-5,1′-cycloalkanes], and evaluation of their analgesic effects in a mouse pain model." *Bioorganic & Medicinal Chemistry Letters 27*, no. 7 (2017): 1551-1556. Accessed July 9, 2023. https://doi.org/10.1016/j.bmcl.2017.02.036.

Cai, Xiaoying, Wang, Lun, Yi, Yuyao, Deng, Dexin, Shi, Mingsong, Tang, Minghai, Li, Na, Wei, Haoche, Zhang, Ruijia, Su, Kaiyue, Ye, Haoyu, Ch, Lijuan. "Discovery of pyrimidine-5-carboxamide derivatives as novel salt-inducible kinases (SIKs) inhibitors for inflammatory bowel disease (IBD) treatment." *European Journal of Medicinal Chemistry 256*, (2023): 115469. Accessed July 11, 2023. https://doi.org/10.1016/j.ejmech.2023.115469.

Caique Santos Costa, Érick, Zenaide S. Monte, Emerson P. S. Falcão, Mayara dos Santos Maia, Renata Priscila Barros de Menezes, Luciana Scotti, Marcus Tullius Scotti, and Sebastião José de Melo. 2023. "Pharmacological Potential of Pyrmidine Derivatives: A Review with Emphasis on Antiviral Effects and Virtual Screening against Sars-Cov-2 Molecular Targets." *Chemistry Select* 8 (23). https://doi.org/10.1002/ slct.202300132.

Chen, Linlin, Huidan Deng, Hengmin Cui, Jing Fang, Zhicai Zuo, Junliang Deng, Yinglun Li, Xun Wang, and Ling Zhao. 2018. "Inflammatory Responses and Inflammation-Associated Diseases in Organs." *Oncotarget* 9 (6): 7204–18. https://doi.org/10.18632/oncotarget.23208.

Chkirate, Karim, Fettach, Saad, El Hafi, Mohamed, Karrouchi, Khalid, Elotmani, Bouchaib, Mague, Joel T., Radi, Smaail, Faouzi, My El Abbes, Adarsh, N. N., Essassi, El Mokhtar, Garcia, Yann. "Solvent induced supramolecular polymorphism in Cu(II) coordination complex built from 1,2,4-triazolo[1,5-a]pyrimidine: Crystal structures and anti-oxidant activity." *Journal of Inorganic Biochemistry 208*, (2020): 111092. Accessed July 9, 2023. https://doi.org/10.1016/j.jinorgbio.2020.111092.

Dave, Shikha S., and Anjali M. Rahatgaonkar. "Syntheses and anti-microbial evaluation of new quinoline scaffold derived pyrimidine derivatives." *Arabian Journal of Chemistry 9*, (2016): S451-S456. Accessed July 9, 2023. https://doi.org/10.1016/j.arabjc.2011.06.009.

Day, Richard O, and Garry G Graham. 2004. "The Vascular Effects of COX-2 Selective Inhibitors." *Australian Prescriber* 27 (6): 142–45. https://doi.org/10.18773/austprescr.2004.119.

Kirillov, Ivan A., Nawrozkij, Maxim B., Vernigora, Andrey A., Brunilina, Leila L., Fiorentino, Francesco, Mladenović, Milan, Rotili, Dante, Ragno, Rino. "Pyrimidine thioethers: A novel class of antidepressant agents, endowed with anxiolytic, performance enhancing and nootropic activity." *European Journal of Medicinal Chemistry 245*, (2023): 114902. Accessed July 10, 2023. https://doi.org/10.1016/j.ejmech.2022.114902.

Hanna, Mona M. "New pyrimido[5,4-e]pyrrolo[1,2-c]pyrimidines: Synthesis, 2D-QSAR, anti-inflammatory, analgesic and ulcerogenicity studies." *European Journal of Medicinal Chemistry 55*, (2012): 12-22. Accessed July 9, 2023. https://doi.org/10.1016/j.ejmech.2012.06.048.

Hijos-Mallada, Gonzalo, Sostres, Carlos, and Fernando Gomollón. "NSAIDs, gastrointestinal toxicity and inflammatory bowel disease." *Gastroenterología y Hepatología (English Edition) 45*, no. 3 (2022): 215-222. Accessed July 10, 2023. https://doi.org/10.1016/j.gastre.2021.06.002.

Kantankar, Abhijit, Jayaprakash Rao, Y., Mallikarjun, G., Hemasri, Y., and Raghava R. Kethiri. "Rational design, synthesis, biological evaluation and molecular docking studies of chromone-pyrimidine derivatives as potent anti-cancer agents." *Journal of Molecular Structure 1239*, (2021): 130502. Accessed July 9, 2023. https://doi.org/10.1016/j.molstruc.2021.130502.

Keri, Rangappa S., Hosamani, Kallappa M., Shingalapur, Ramya V., and Mallinath H. Hugar. "Analgesic, anti-pyretic and DNA cleavage studies of novel pyrimidine derivatives of coumarin moiety." *European Journal of Medicinal Chemistry 45*, no. 6 (2010): 2597-2605. Accessed July 9, 2023. https://doi.org/10.1016/j.ejmech.2010.02.048.

Khazimullina, Y. Z., A. R. Gimadieva, V. R. Khairullina, L. F. Zainullina, Y. V. Vakhitova, and A. G. Mustafin. 2022. "The Synthesis and Anti-Inflammatory Studies of New Pyrimidine Derivatives, Inhibitors of Cyclooxygenase Isoforms." *Russian Journal of Bioorganic Chemistry* 48 (5): 1027–35. https://doi.org/10.1134/s1068162022050107.

Khoo, Boyuan, Crene, Elliot, Le, Dianne, and Wayne Ng. "Perioperative NSAID use in single level microdiscectomy and hemilaminectomy." *Interdisciplinary Neurosurgery 31*, (2023): 101679. Accessed July 10, 2023. https://doi.org/10.1016/j.inat.2022.101679.

Kim, Hee J., Kim, Dong, and Sung H. Um. "The Novel Inhibitory Effect of YM976 on Adipocyte Differentiation." *Cells 12*, no. 2 (2023): 205. Accessed July 11, 2023. https://doi.org/10.3390/cells12020205.

Lagoja, Irene M. 2005. "Pyrimidine as Constituent of Natural Biologically Active Compounds." *Chemistry & Biodiversity* 2 (1): 1–50. https://doi.org/10.1002/cbdv.200490173.

Löffler, M., and E. Zameitat. 2013. "Pyrimidine Biosynthesis and Degradation (Catabolism)." Edited by William J. Lennarz and M. Daniel Lane. ScienceDirect. Waltham: Academic Press. January 1, 2013. https://www.sciencedirect.com/science/article/abs/pii/B978012378630200178X.

Manzoor, Shoaib, Daniyah A. Almarghalani, Antonisamy William James, Md Kausar Raza, Tasneem Kausar, Shahid M. Nayeem, Nasimul Hoda, and Zahoor A. Shah. 2022. "Synthesis and Pharmacological Evaluation of Novel Triazole-Pyrimidine Hybrids as Potential Neuroprotective and Anti-Neuroinflammatory Agents." *Pharmaceutical Research* 40 (1). https://doi.org/10.1007/s11095-022-03429-1.

Medzhitov, Ruslan. "Origin and physiological roles of inflammation." *Nature 454*, no. 7203 (2008): 428-435. Accessed July 10, 2023. https://doi.org/10.1038/nature07201.

Meneghesso, Silvia, Vanderlinden, Evelien, Stevaert, Annelies, McGuigan, Christopher, Balzarini, Jan, and Lieve Naesens. "Synthesis and biological evaluation of pyrimidine nucleoside monophosphate prodrugs targeted against influenza virus." *Antiviral Research 94*, no. 1 (2012): 35-43. Accessed July 9, 2023. https://doi.org/10.1016/j.antiviral.2012.01.007.

Morsy, Nesrin M., Khadiga M. Abu-Zied, Ahmed I. Saleh, and Abdelbaset M. Elgamal. 2022. "Design, Synthesis, Molecular Docking of Some New Polyhydrobenzothieno-Thiazolopyrimidinedione Glycoside Derivatives with Double Anti-Microbial-Anti-Inflammatory Action." *Egyptian Journal of Chemistry* 0 (0). https://doi.org/10.21608/ejchem.2022.155556.6714.

Mushtaq, Alia, Azam, Uzma, Mehreen, Saba, and Muhammad M. Naseer. "Synthetic α-glucosidase inhibitors as promising anti-diabetic agents: Recent developments and future challenges." *European Journal of Medicinal Chemistry 249*, (2023): 115119. Accessed July 10, 2023. https://doi.org/10.1016/j.ejmech.2023.115119.

Myriagkou, Malama, Papakonstantinou, Evangelia, Deligiannidou, Georgia, Patsilinakos, Alexandros, Kontogiorgis, Christos, and Eleni Pontiki. "Novel Pyrimidine Derivatives as Antioxidant and Anticancer Agents: Design, Synthesis and Molecular Modeling Studies." *Molecules 28*, no. 9 (2023): 3913. Accessed July 11, 2023. https://doi.org/10.3390/molecules28093913.

Nagalakshmamma, Vadabingi, Venkataswamy, Mallepogu, Pasala, Chiranjeevi, Umamaheswari, Amineni, Thyagaraju, Kedam, Nagaraju, Chamarthi, and Ponne V. Chalapathi. "Design, synthesis, anti-tobacco mosaic viral and molecule docking simulations of urea/thiourea derivatives of 2-(piperazine-1-yl)-pyrimidine and 1-(4-Fluoro/4-Chloro phenyl)-piperazine and 1-(4-Chloro phenyl)-piperazine – A study." *Bioorganic Chemistry 102*, (2020): 104084. Accessed July 9, 2023. https://doi.org/10.1016/j.bioorg.2020.104084.

Nielsen, O H., Andersen, P S., and S E. Girardin. "Chronic inflammation: importance of NOD2 and NALP3 in interleukin-1β generation." *Clinical and Experimental Immunology 147*, no. 2 (2007): 227-235. Accessed July 10, 2023. https://doi.org/10.1111/j.1365-2249.2006.03261.x.

Panchal, Nagesh K., and Evan Prince Sabina. "Non-steroidal anti-inflammatory drugs (NSAIDs): A current insight into its molecular mechanism eliciting organ toxicities." *Food and Chemical Toxicology 172*, (2023): 113598. Accessed July 10, 2023. https://doi.org/10.1016/j.fct.2022.113598.

Patil, Sharanabasappa B. "Recent medicinal approaches of novel pyrimidine analogs: A review." *Heliyon 9*, no. 6 (2023): e16773. Accessed July 9, 2023. https://doi.org/10.1016/j.heliyon.2023.e16773.

Peytam, Fariba, Takalloobanafshi, Ghazaleh, Saadattalab, Toktam, Norouzbahari, Maryam, Emamgholipour, Zahra, Moghimi, Setareh, Firoozpour, Loghman, Bijanzadeh, Hamid Reza, Faramarzi, Mohammad Ali, Mojtabavi, Somayeh, Rashidi-Ranjbar, Parviz, Karima, Saeed, Pakraad, Roya and Foroumadi, Alireza. "Design, synthesis, molecular docking, and in vitro α-glucosidase inhibitory activities of novel 3-amino-2,4-diarylbenzo[4,5]imidazo[1,2-a]pyrimidines against yeast and rat α-glucosidase." *Scientific Reports 11*, no. 1 (2021): 1-18. Accessed July 9, 2023. https://doi.org/10.1038/s41598-021-91473-z.

Rambabu, G., Y. B. Kiran, S. Sarveswari, and V. Vijayakumar. 2021. "Synthesis of New 6-Oxo-1,6-Dihydropyrimidin-5-Carboxamides and Evaluation of Their Anti-Bacterial and Anti-Oxidant Activities." *Polycyclic Aromatic Compounds* 43 (1): 619–29. https://doi.org/10.1080/10406638.2021.2019799.

Rani, Jyoti, Sanjiv Kumar, Monika Saini, Jyoti Mundlia, and Prabhakar Kumar Verma. 2016. "Biological Potential of Pyrimidine Derivatives in a New Era." *Research on Chemical Intermediates* 42 (9): 6777–6804. https://doi.org/10.1007/s11164-016-2525-8.

Rashid, Haroon ur, Marco Antonio Utrera Martines, Adriana Pereira Duarte, Juliana Jorge, Shagufta Rasool, Riaz Muhammad, Nasir Ahmad, and Muhammad Naveed Umar. 2021. "Research Developments in the Syntheses, Anti-Inflammatory Activities and Structure–Activity Relationships of Pyrimidines." *RSC Advances* 11 (11): 6060–98. https://doi.org/10.1039/d0ra10657g.

Rashid, Haroon ur, Yiming Xu, Nasir Ahmad, Yaseen Muhammad, and Lisheng Wang. 2019. "Promising Anti-Inflammatory Effects of Chalcones via Inhibition of Cyclooxygenase, Prostaglandin E2, Inducible NO Synthase and Nuclear Factor Kb Activities." *Bioorganic Chemistry* 87 (June): 335–65. https://doi.org/10.1016/j.bioorg.2019.03.033.

Sayed, Amira I., Yara E. Mansour, Mohamed A. Ali, Omnia Aly, Zainab M. Khoder, Ahmed S. Said, Samar S. Fatahala, and Rania H. Abd El-Hameed. 2022. "Novel Pyrrolopyrimidine Derivatives: Design, Synthesis, Molecular Docking, Molecular Simulations and Biological Evaluations as Antioxidant and Anti-Inflammatory Agents." *Journal of Enzyme Inhibition and Medicinal Chemistry* 37 (1): 1821–37. https://doi.org/10.1080/14756366.2022.2090546.

Singh, Karanvir, Pal, Rohit, Khan, Shah A., Kumar, Bhupinder, and Md J. Akhtar. "Insights into the structure activity relationship of nitrogen-containing heterocyclics for the development of antidepressant compounds: An updated review." *Journal of Molecular Structure 1237*, (2021): 130369. Accessed July 10, 2023. https://doi.org/10.1016/j.molstruc.2021.130369.

Sivagami, Swaminathan, Kavitha, Rengarajan, Satanathan, Sasikurba, Rajesh, Jegathalaprathaban, Narenkumar, Jayaraman, Parthipan, Punniyakotti, Muthusamy, Karnan, and Ahmed Alfarhan. "Multicomponent one-pot synthesis, characterization and antimicrobial screening of 2 cyanoimino-6-aryl-4-(6-methoxynaphthalen-2-yl)-3,4-dihydro-1H-pyrimidines." *Process Biochemistry 123*, (2022): 63-69. Accessed July 9, 2023. https://doi.org/10.1016/j.procbio.2022.10.032.

Spasov, Alexander, Ovchinnikova, Irina, Fedorova, Olga, Titova, Yulia, Babkov, Denis, Kosolapov, Vadim, Borisov, Alexander, Sokolova, Elena, Klochkov, Vladlen, Skripka, Maria, Velikorodnaya, Yulia, Smirnov, Alexey, Rusinov, Gennady, and Charushin, Valery. "Amino Derivatives of Diaryl Pyrimidines and Azolopyrimidines as Protective Agents against LPS-Induced Acute Lung Injury." *Molecules 28*, no. 2 (2023): 741. Accessed July 11, 2023. https://doi.org/10.3390/molecules28020741.

Sun, Duxin, Wei Gao, Hongxiang Hu, and Simon Zhou. 2022. "Why 90% of Clinical Drug Development Fails and How to Improve It?" *Acta Pharmaceutica Sinica B* 12 (7). https://doi.org/10.1016/j.apsb.2022.02.002.

Tay, Sen H., Santosa, Amelia, Goh, Eugene C. H., Xu, Chun X., Wu, Lik H., Bigliardi-Qi, Mei, Pakkiri, Leroy S. S., Lee, Bernett T. K., Drum, Chester L., and Paul L. Bigliardi. "Distinct transcriptomic and metabolomic profiles characterize NSAID-induced urticaria/angioedema patients undergoing aspirin desensitization." *Journal of Allergy and Clinical Immunology 150*, no. 6 (2022): 1486-1497. Accessed July 10, 2023. https://doi.org/10.1016/j.jaci.2022.07.025.

Tomma, Jumbad H., Mustafa S. Khazaal, and Ammar H. Al-Dujaili. 2014. "Synthesis and Characterization of Novel Schiff Bases Containing Pyrimidine Unit." *Arabian Journal of Chemistry* 7 (1): 157–63. https://doi.org/10.1016/j.arabjc.2013.08.024.

Veeranna, Navaneethgowda P., Yadav D. Bodke, Manjunatha Basavaraju, and Pruthviraj Krishnamurthy. 2022. "An Efficient One Pot Synthesis of (2-Hydroxyphenyl)-5-Phenyl-6-(Pyrrolidine-1-Carbonyl)-1H-Pyrano[2,3-d]Pyrimidine-2,4(3H,5H)-Dione Derivatives as a Potent Biological Agents." *Nucleosides, Nucleotides & Nucleic Acids* 42 (4): 281–95. https://doi.org/10.1080/15257770.2022.2127765.

Wang, Hongwei, Cui, Enjing, Li, Jiaming, Ma, Xiaodong, Jiang, Xueyang, Du, Shuaishuai, Qian, Shihu, and Le Du. "Design and synthesis of novel indole and indazole-piperazine pyrimidine derivatives with anti-inflammatory and neuroprotective activities for ischemic stroke treatment." *European Journal of Medicinal Chemistry 241*, (2022): 114597. Accessed July 11, 2023. https://doi.org/10.1016/j.ejmech.2022.114597.

Wang, Hongwei, Cui, Enjing, Li, Jiaming, Ma, Xiaodong, Jiang, Xueyang, Du, Shuaishuai, Qian, Shihu, and Le Du. "Design and synthesis of novel indole and indazole-piperazine pyrimidine derivatives with anti-inflammatory and neuroprotective activities for ischemic stroke treatment." *European Journal of Medicinal Chemistry 241*, (2022): 114597. Accessed July 9, 2023. https://doi.org/10.1016/j.ejmech.2022.114597.

Wang, Yilin, Wu, Jin, Shen, Ruihan, Li, Yubao, Ma, Guofeng, Qi, Shuang, Wu, Wenjuan, Jin, Yongcan, and Bo Jiang. "A mild iodocyclohexane demethylation for highly enhancing antioxidant activity of lignin." *Journal of Bioresources and Bioproducts*, (2023). Accessed July 10, 2023. https://doi.org/10.1016/j.jobab.2023.05.001.

Zarenezhad, Elham, Mojtaba Farjam, and Aida Iraji. 2021. "Synthesis and Biological Activity of Pyrimidines-Containing Hybrids: Focusing on Pharmacological Application." *Journal of Molecular Structure* 1230 (April): 129833. https://doi.org/10.1016/j.molstruc.2020.129833.

Chapter 2

# Studies on Pyrimidine Derivatives

**Hüseyin Fatih Çetinkaya[1]**
**Muhammed Safa Çelik[2]**
**Nübar Abbaszada[2]**
**Serap Çetinkaya[2]**
**and Burak Tüzün[3,*]**

[1]Sivas Cumhuriyet University, Faculty of Engineering, Department of Environmental Engineering, Sivas, Turkey
[2]Sivas Cumhuriyet University, Science Faculty, Department of Molecular Biology and Genetics, Sivas, Turkey
[3]Plant and Animal Production Department, Technical Sciences Vocational School of Sivas, Sivas Cumhuriyet University, Sivas, Turkey

## Abstract

Pyrimidine is a nitrogen-containing heterocyclic aromatic compound. Together with its derivatives it plays an expansive role in drug discovery studies. In the chemistry of biological systems, pyrimidine derivatives have engrossed interests as they formed the bases of therapeutic natural products. They are the building blocks of many natural compounds: antibiotics, liposaccharides, and vitamins, to name a few. Pyrimidine is used in heterocyclic drugs with antimicrobial, anti-inflammatory, antidiabetic, antiviral, anticancer, and antifungal activities. Hence, it also serves as a versatile tool in the theoretical development of heterocyclic chemistry and organic synthesis. This study attempted to cover various pyrimidine derivatives with powerful biological and pharmacological applications.

* Corresponding Author's E-mail: theburaktuzun@yahoo.com.

In: The Chemistry of Pyrimidine Derivatives
Editor: Barry Schneider
ISBN: 979-8-89113-563-5

**Keywords:** anticancer, antimicrobial, anti-inflammatory, antiviral activity, medicinal drugs pyrimidine

## 1. Introduction

Pyrimidine is a benzene and contains two nitrogen atoms. It is one of the three diazine isomers (Figure 1). Purines also include a pyrimidine in their structure as well as a fused imidazole ring.

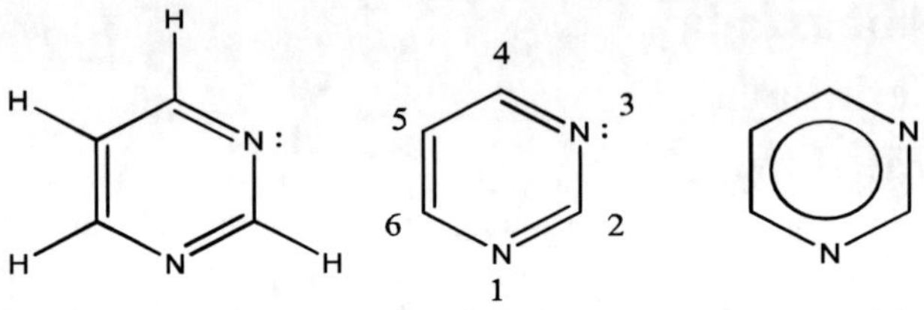
H
H
N:
H
N
H
4
5
3
N:
6
2
N
1
N
N

**Figure 1.** Pyrimidine and its two isomers.

Purines and pyrimidines are known as the five bases of nucleic acid polymers, including both ribo-and deoxyribonucleotides, RNA and DNA, respectively. Together with their substituted forms and tautomers, they form an abundant group of organic material in nature. Nucleic acid polymers can extend as single-or double helices by the inherent virtue of complementarity between purines and pyrimidines through hydrogen bonding [2].

Pyrimidine has recently entered in the spacious realm of pure and applied chemistry. Its core in its molecular structure has been the subject of many research efforts [3-4]. A wide range of functional groups can bind around this core and enrich its environmental friendly functional properties [5-9]. Especially, heterocyclic compounds have been produced and their biological properties have been assessed [10-14]. To this end, a cohort of medicines has been derived from three pyrimidines, thymine, cytosine, and uracil, derivatives. Pyrimidine analogues have been exploited as anticoagulation agents and anticancer drugs, 5-fluorouracil, for example, contraceptives, and in the treatment of Parkinson disease [15]. Idoxuridine and trifluoridine, zidovudine and stavudine are four of the known antiviral agents. Three microbial drugs, sulfamethazine and sulfadiazine, sulfadoxine have also been derived from pyrimidines. The latter can also be used as an anti-malarial agent. Other derivatives minoxidil and prazosin are used for the treatment of hypertension. Sedatives include another derivative, phenobarbitone [16].

## 2. Clinical Significance of Pyrimidine Analogues

Quite a few chemotherapeutic drugs have been synthesized as pyrimidine analogues [17-19] and nucleotide analogues have been produced to be used as nucleosides, nucleotides and coenzymes [20].

### 2.1. Anti-Cancerous Pyrimidine Derivatives

Cancer is a multifactorial disease that is widely recognized as the world's most serious health problem. Despite recent advances in our understanding of the biological mechanisms that contribute to cancer growth. MTT assays were used to test various pyrimidine-bridged combretastatin derivatives for anticancer activity against breast cancer (MCF-7) and lung cancer (A549) cell lines. MTT is a colorimetric test used for the measurement of cell proliferation. Most of the synthesized compounds with IC50 values in the low micromolar band showed potent anticancer activity [21]. Cancer is the deadliest disease and the second leading cause of death in developing countries. According to the International Agency for Research on Cancer, 14.1 million cancer cases were recorded in 2012 and 8.2 million cancer patients died [16, 22].

In the study, it was reported that a new line of 1,2,4-oxadiazole-linked 4-(oxazolo[5,4-d] pyrimidine derivatives was designed, synthesized and tested for anticancer activity against four human cancer cell lines such as breast cancer (Figure 2). Breast cancer (MCF-7), lung cancer (A-549), colon cancer (Colo-205) and ovarian cancer (A2780) The majority of compounds screened exhibited moderate to excellent anticancer activity against all cell lines examined [16].

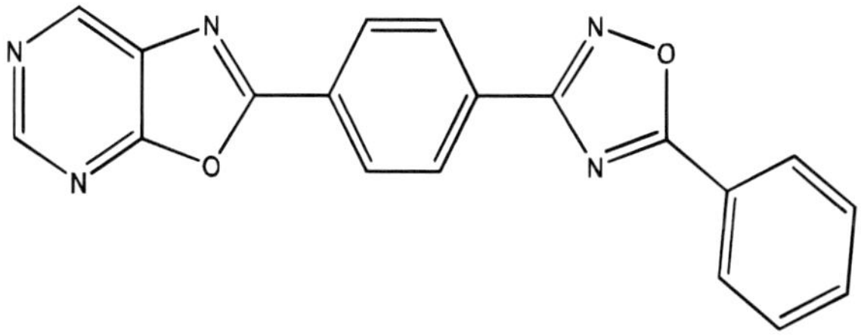
N
N
N
O
N
O
N

**Figure 2.** Structure of 1,2,4-oxadiazole linked 4-(oxazolo[5,4-d] pyramiding.

Antifolates endorse inhibition of replication, doubling of genomic DNA, in cancer cells [23-26]. They do this prevention by competing with the cell's own metabolites on account of structural similarities to them. Resulting inhibition often leads the cell to apoptotic pathways [27]. Folate-antagonists, miming folates, 2-amino-4-hydroxypyrimidines and 2,4 diaminopyrimidines, move into the cell and hinder the function of dihydrofolate reductase [28]. The role of dihydrofolate reductase resides in the metabolic pathways involving amino acid- and purine nucleotide synthesis [29]. Methotrexate (1), for example, a folate antagonist, has been exploited in the treatment of rigid tumours and malignant liver cancers (Figure 3) [30-32].

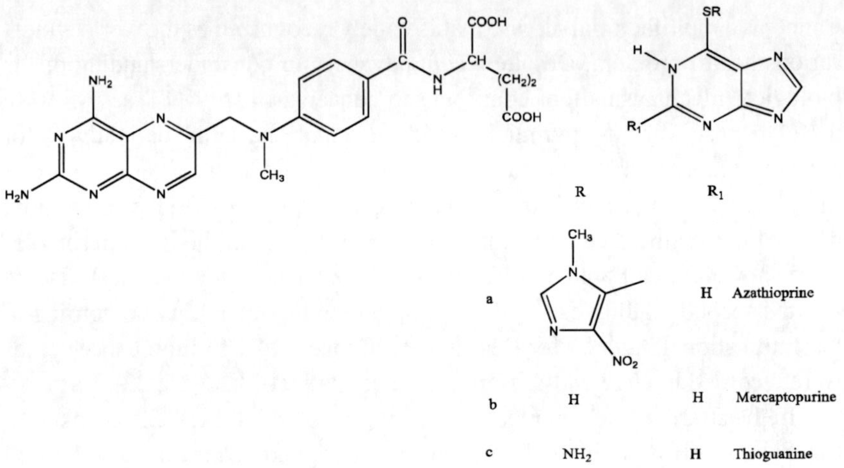
NH2
H2N
CH3
COOH
(CH2)2
COOH
SR
R1
R
R1
a
H Azathioprine
NO2
b
H
H Mercaptopurine
c
NH2
H Thioguanine

**Figure 3.** Folate-antagonists as anticancer drugs.

Two purine pyrimidine fusion, azathioprine (2a), 6-mercaptopurine (2b), and 6-thioguanine (2c) are also among the anti-cancer drugs [33-35]. Pyrimidine-antagonists, miming pyrimidines are involved in the prevention of DNA synthesis, thus inhibiting cell proliferation [36]. For example, 5-fluorouracil (3), 5-thiouracil (4) and tegafur (5) all exert anticancer activities [37-39]. In addition, a cytosine nucleoside, gemcitabine (6), is also an anti-cancer meaningfully effective against a number of cancer cell lines [40]. Mopidamol (7), nimustine (8), uramustine (9), trimetrexate glucuronate (10), and 1-β-D-arabinosylcytosine (Ara-C) (11) also include a pyrimidine moiety in their structure (Figure 4) [32].

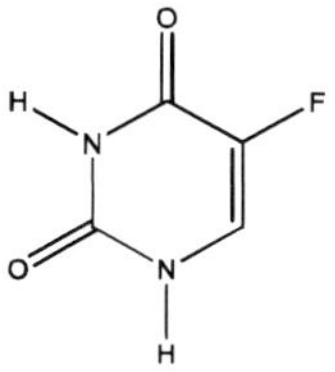
3

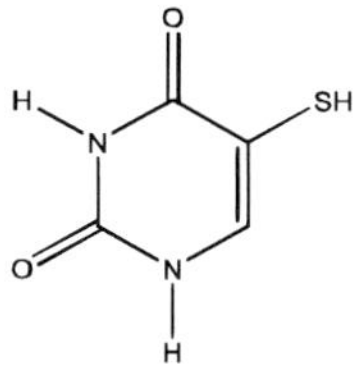
4

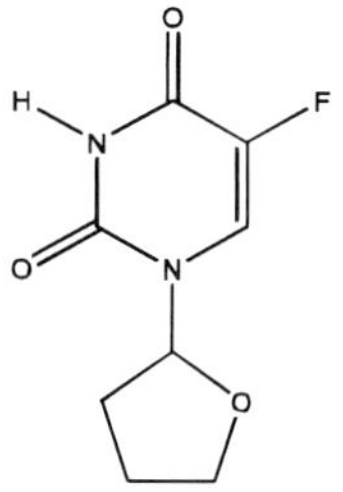
5

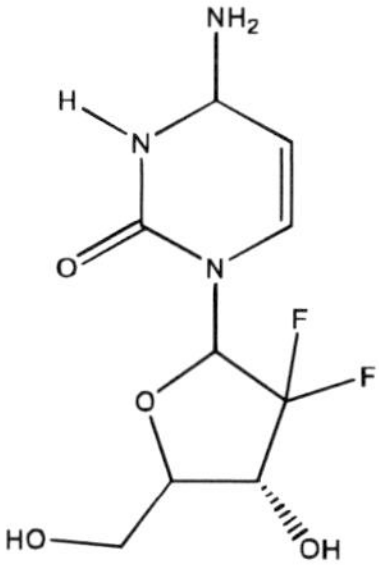
6

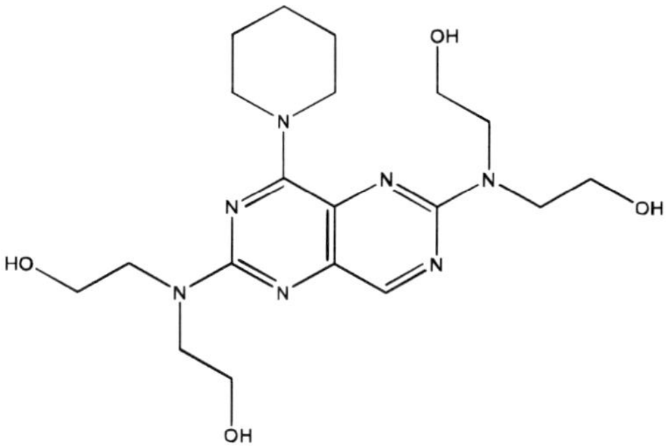
7

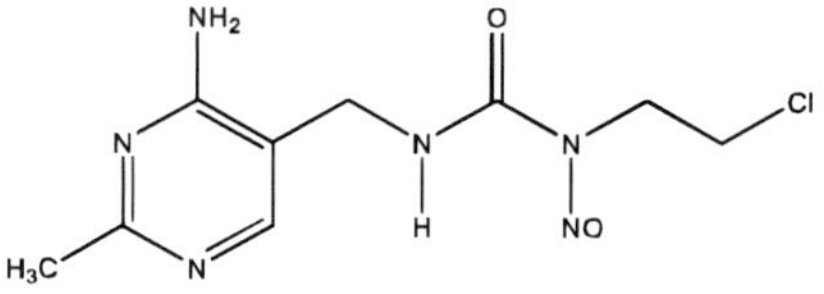
8

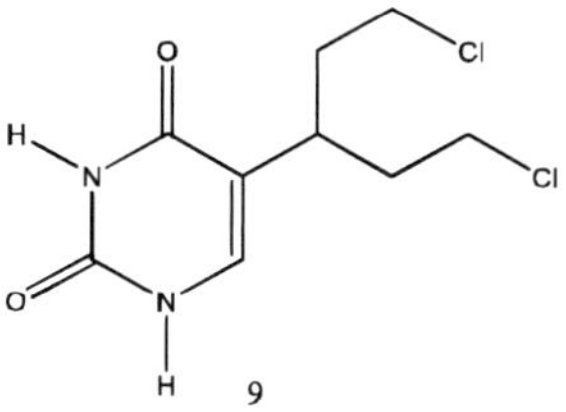
9

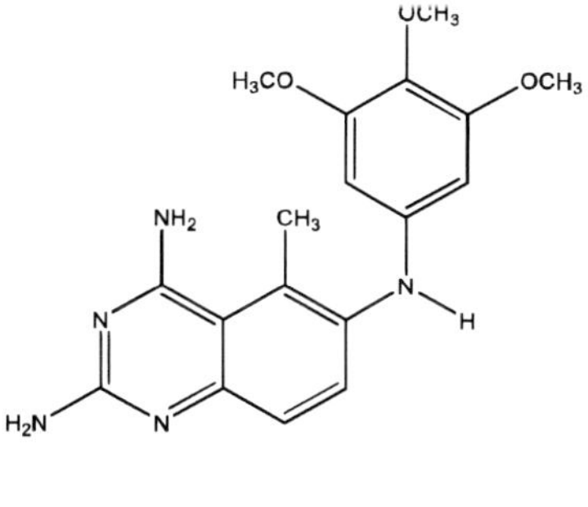
10

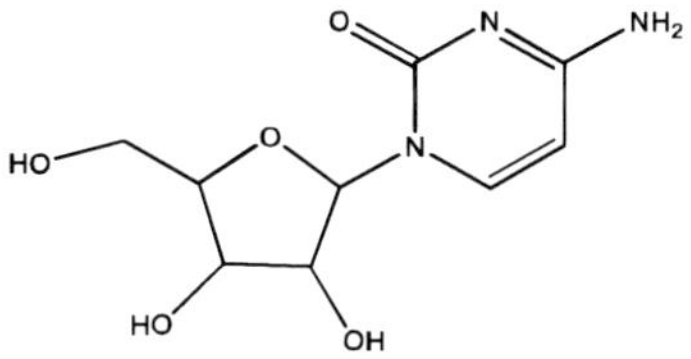
11

**Figure 4.** Structures of pyrimidine analogues as anticancer drugs.

## 2.2. Sedative and Hypnotic Pyrimidine Analogues

The core of barbiturates is often made up of barbituric acid, and they are commercially available medicines, causing sedation or anaesthesia in the central nervous system [41, 42], while barbituric acid per se does not have such a property. In addition, allobarbital (12) and aprobarbital (13) are hypnotic agents. Pentobarbital (14), another barbiturate, is another sedative applied in the cases of emergency [43]. Anti-insomniac agents include cyclobarbital and/or cyclobarbitol (15), and diazepam [44]. Propallylonal (16), ibomal, also known as nostal, and phenobarbital are useful for their sedative, hypnotic, and anticonvulsant activities (Figure 5) [32, 45, 46] (17).

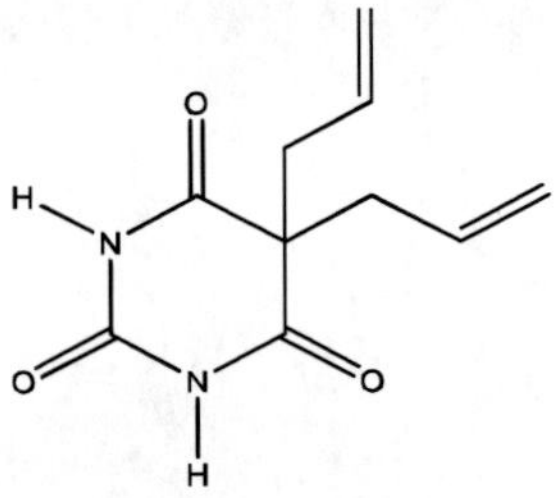

12

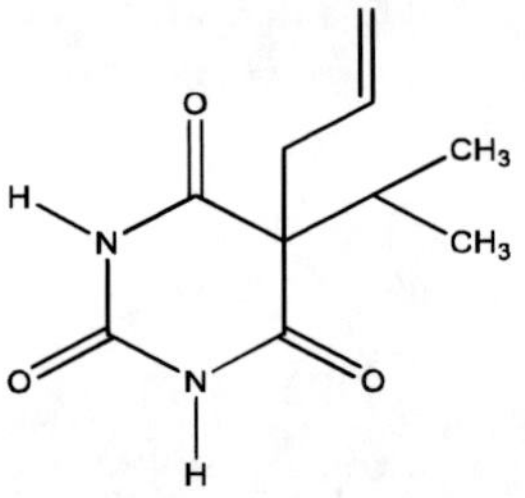

13

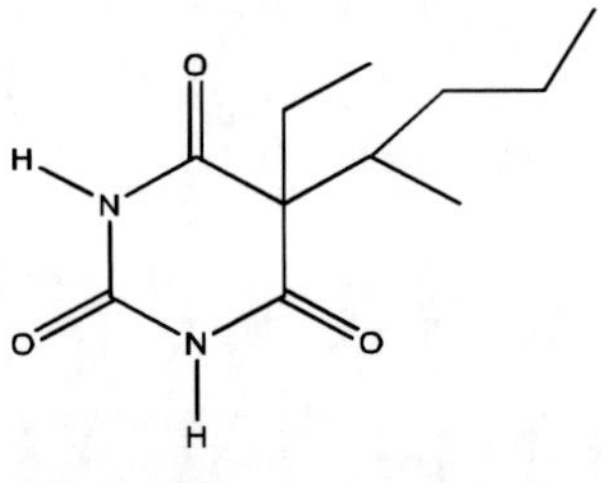

14

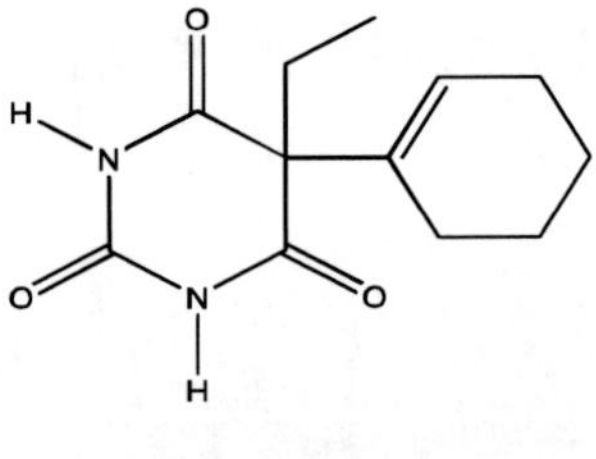

15

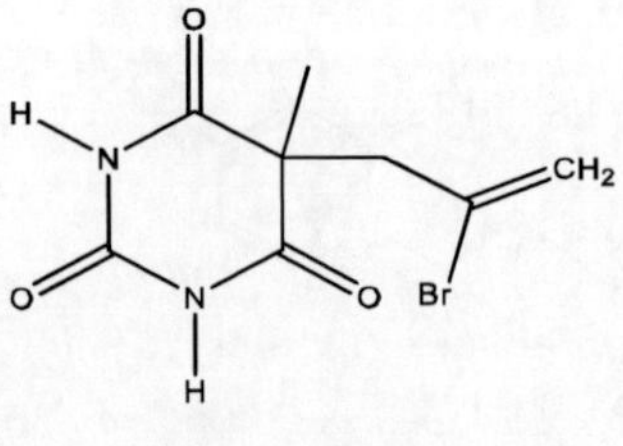

16

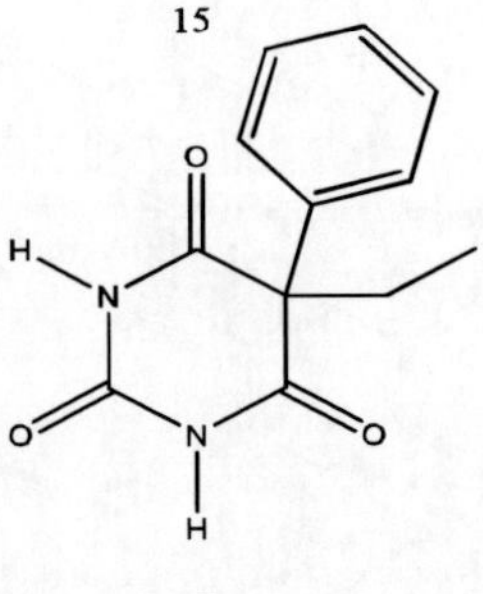

17

**Figure 5.** Structures of sedative and hypnotic pyrimidine analogues.

## 2.3. Antibiotic Pyrimidine Derivatives

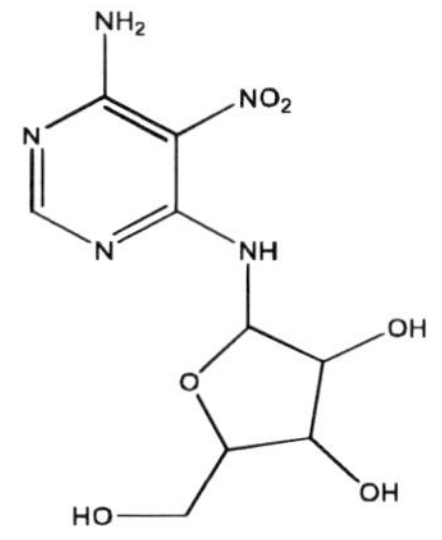

18

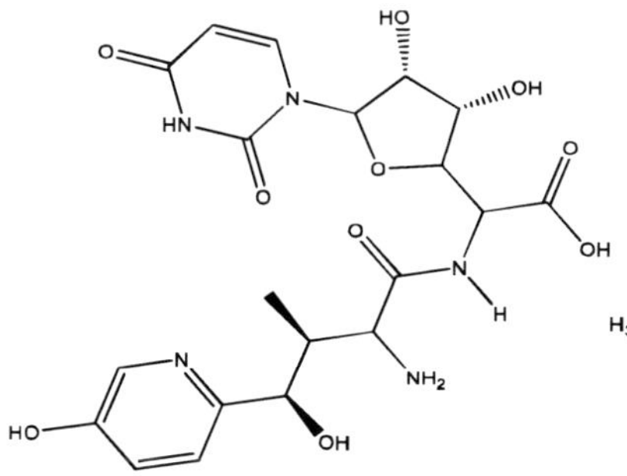

19

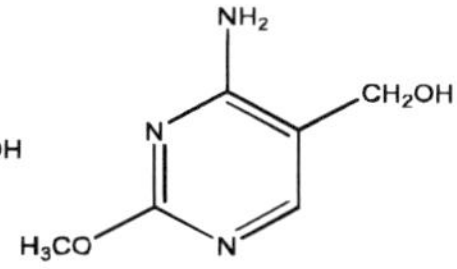

20

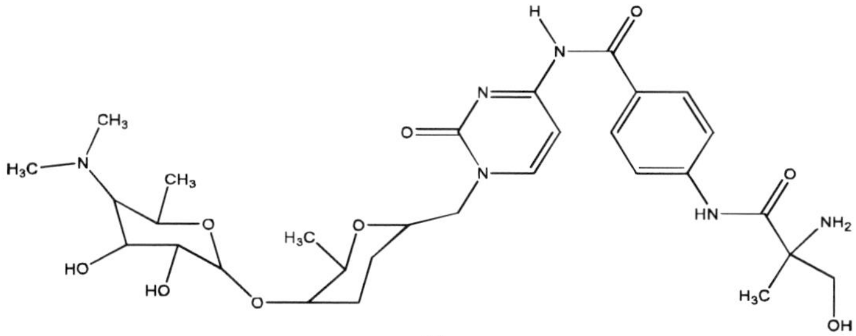

21

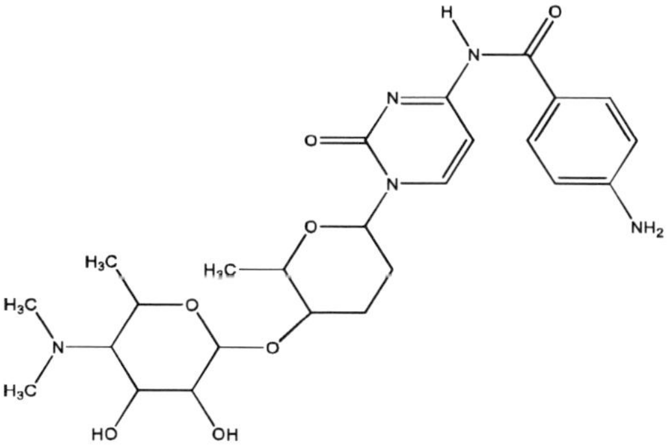

22

**Figure 6.** Structures of antibiotic pyrimidine analogues.

Pyrimidine nucleoside analogues appear to have a wide range of cidal activities. Some prominent examples were cited here. Clitosine (18), harbours insecticidal- and cytostatic activities in some of the leukaemia cell lines [47]. Nikkomycin Z (19), has antifungal activity [48] (Figure 6). An antibiotic, basimethrin [5-hydroxymethyl-2-methoxypyrimidine-4 amine] (20), is effective on some bacterial species [49]. Amicitin (21) and plicacetin (22) are the antibiotics used against Gram-positive bacteria [32, 50, 51].

Fused pyrimidine derivatives, especially pyrido[2,3-d]pyrimidine derivatives, have attracted much attention among pyrimidine-containing compounds as they have interesting bioactivities, including antifungal activity [52]. An outstanding example of these has been pyrazolo[4,3:5,6]pyrido[2,3-d]pyrimidine (Figure 7). Its synthesis involved a microwave-assisted reaction. This method has been rather innovative and efficient in terms of both application and yield [16, 52].

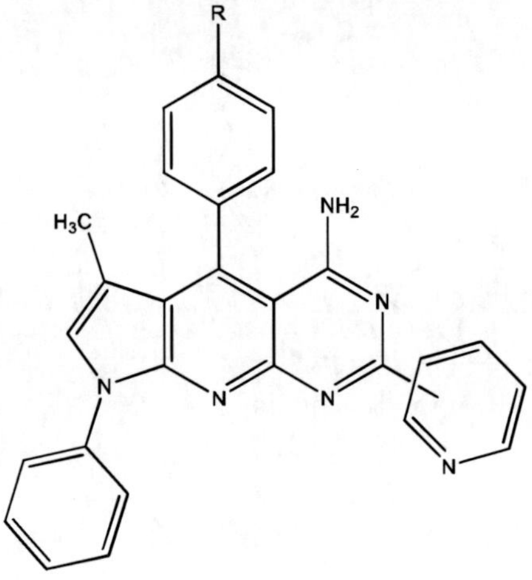
R
$H_3C$
$NH_2$
N
N
N
N
N

**Figure 7.** Structure of pyrazolo[4,3:5,6]pyrido[2,3-d]pyrimidines.

## 2.4. Antiviral Pyrimidine Derivatives

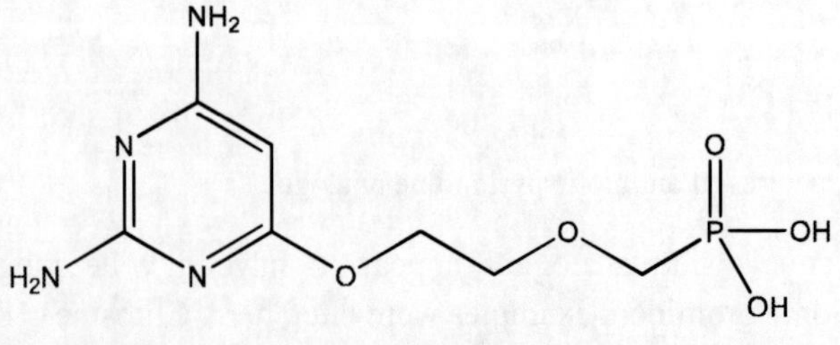
$NH_2$
N
$H_2N$
N
O
O
O
P
OH
OH

**Figure 8.** Structure of 2,4-diamino-6-[2-(phosphonomethoxy)ethoxy]pyrimidine.

Acyclic nucleoside phosphonates (ANPs) also have important antiviral activities. Two examples of these are 2,4-diamino-6-[2-(phosphonomethoxy) ethoxy]pyrimidines (PMEO-DAPy) pro-fraction containing carbonyloxy-methyl esters (POM, POC), alkoxyalkyl esters, amino acid phosphoramidates and/or tyrosine (Figure 8), and 1-[2-(phosphonome-thoxy)ethyl]-5-azacytosine (PME-5-azaC) (Figure 9) [16, 53].

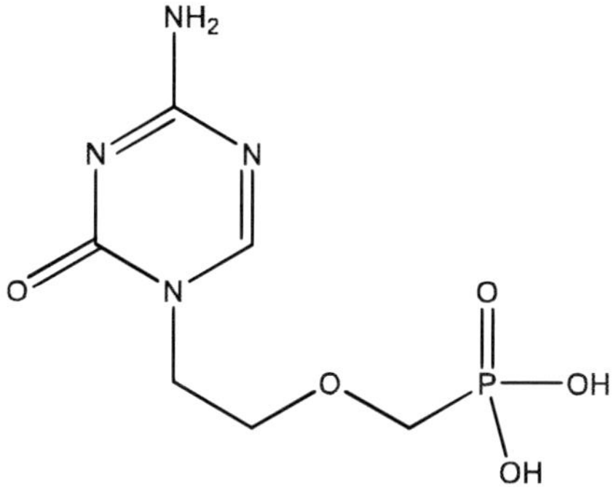
NH2
N
N
N
O
O
O
P
OH
OH

**Figure 9.** Structure of 1-[2-(phosphonomethoxy)ethyl]-5-azacytosine.

### *2.4.1. Anti-Herpes*

Cells invaded by viruses possess exclusive characteristics [54]. Herpes simplex virus (HSV) infections in humans for example triggers the expression of a set of specific enzymes, including thymidine (deoxycytidine) kinase, DNA polymerase, 3`-5`-exodeoxyribonuclease, ribonucleotide reductase, and dUTP pyrophosphatase [55-57]. Some anti-herpes pyrimidine analogues specifically target the above-mentioned enzymes: 5-Iodo-2`-deoxyuridine (Idoxuridine), 5-Trifluoromethyl-2`-deoxyuridine, E-5-(2-Bromovinyl)-2` -deoxvuridine, 9-β-D-Arabinofuranosyladenine, 2`-Fluoro-S-iodo-l-β-D-arabinofuranosylcytosine, Acyclovir and Ganciclovir [32].

### *2.4.2. Anti-HIV*

Reverse transcriptase (RT) of HIV has been the key therapeutic target [58], and a few pyrimidine nucleoside analogues have been approved to serve as anti-HIV agents: 3`- azidothymidine (zidovudine) (30), 2`,3`-dideoxycytidine (DDC) (31), and 2`,3`-didehydro-3`-deoxythymidine (d4T) (32); cidofovir (33), lamivudine (34), lopinavir (36), (37), emtricitabine (38), tenofovir alafenamide (39), and zalcitabine (35) [59-61] (Figure 10). Lopinavir is an antiretroviral protease inhibitor [62]. Together with ritonavir it has been tried on COVID-19 (Figure 11) [32, 63].

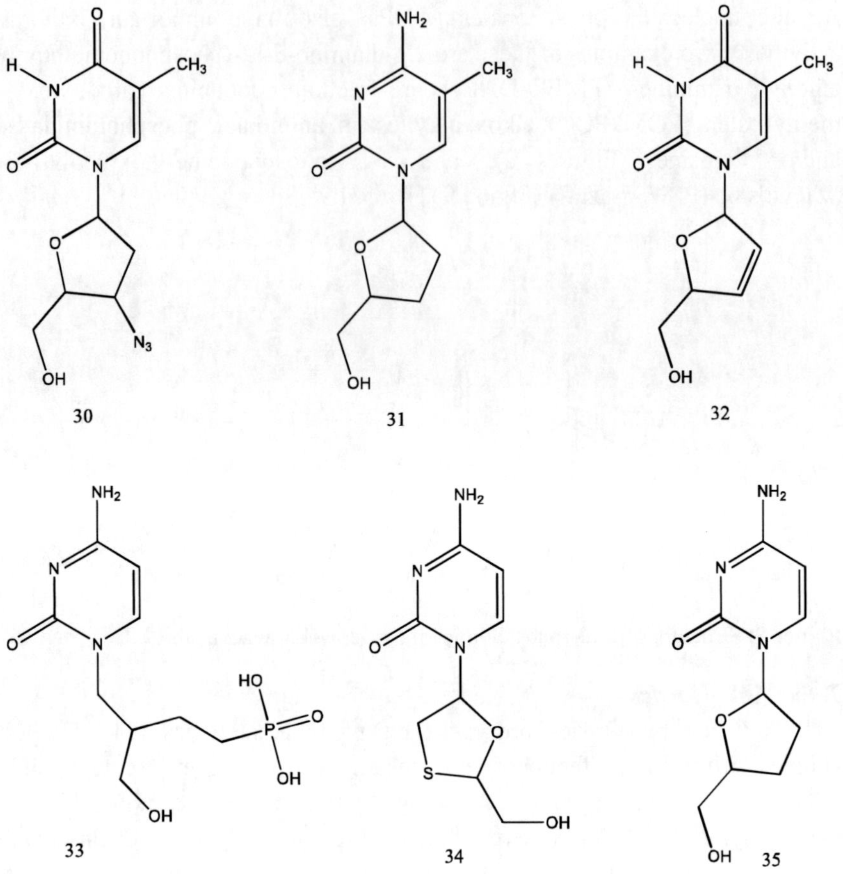
30
31
32
33
34
35

**Figure 10.** Structures of anti-HIV pyrimidine analogues.

## Conclusion

Pyrimidines are normally involved in the make-up of genetic code and RNA. Recently it has been evidenced that their heterocyclic core can also serve the core components in the synthesis of versatile drug molecules. This study tried to introduce the prominent examples of these pyrimidine derivatives that are essentially the anticancer, antimicrobial and antiviral agents.

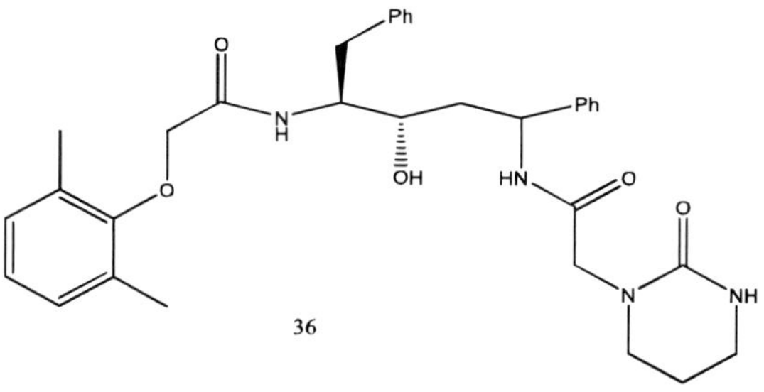
Ph
O
N
H
Ph
O
OH
HN
O
O
N
NH

36

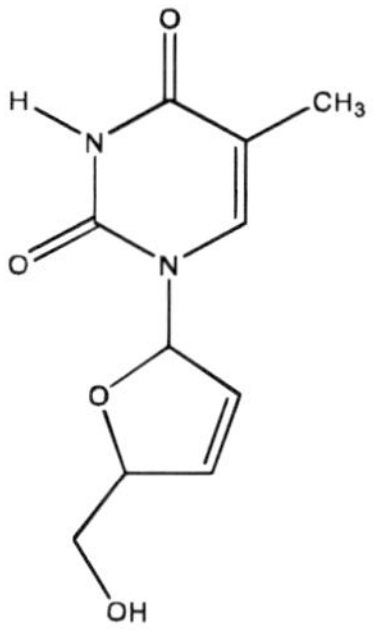
O
H
N
CH3
O
N
O
OH

37

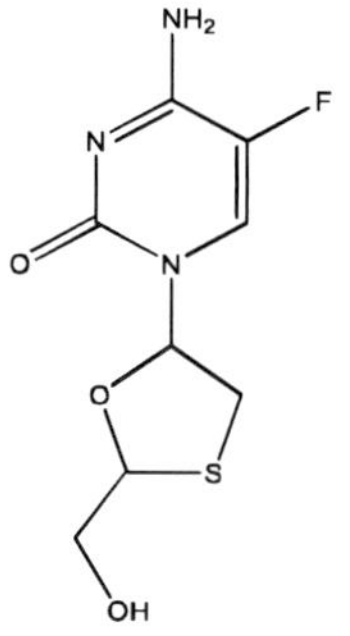
NH2
N
F
O
N
O
S
OH

38

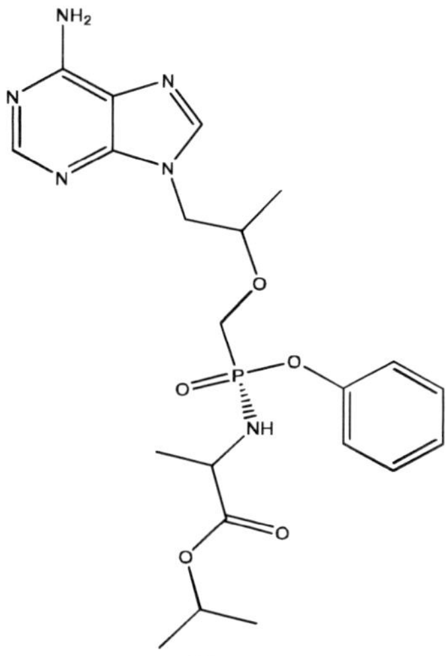
NH2
N
N
N
N
O
O
P
O
NH
O
O

39

**Figure 11.** Structures of anti-HIV pyrimidine analogues.

## References

[1] Patil, S. B. (2018). Biological and medicinal significance of pyrimidines: A review. *Int. J. Pharm. Sci. Res,* 9(1), 44-52.

[2] Sahu, M., & Siddiqui, N. A. D. E. E. M. (2016). A review on biological importance of pyrimidines in the new era. *Int. J. Pharm. Pharm. Sci,* 8(5), 8-21.

[3] Selvam, T. P., James, C. R., Dniandev, P. V., & Valzita, S. K. (2015). A mini review of pyrimidine and fused pyrimidine marketed drugs. *Research in Pharmacy,* 2(4).

[4] Mohana Roopan, S., & Sompalle, R. (2016). Synthetic chemistry of pyrimidines and fused pyrimidines: A review. *Synthetic Communications,* 46(8), 645-672.

[5] Adole, V. A., Waghchaure, R. H., Pathade, S. S., Patil, M. R., Pawar, T. B., & Jagdale, B. S. (2020). Solvent-free grindstone synthesis of four new (E)-7-(arylidene)-indanones and their structural, spectroscopic and quantum chemical

study: a comprehensive theoretical and experimental exploration. *Molecular Simulation,* 46(14), 1045-1054.

[6] Adole, V. A., Jagdale, B. S., Pawar, T. B., & Sagane, A. A. (2020). Ultrasound promoted stereoselective synthesis of 2, 3-dihydrobenzofuran appended chalcones at ambient temperature. *South African Journal of Chemistry,* 73(1), 35-43.

[7] Shinde, R. A., Adole, V. A., Jagdale, B. S., Pawar, T. B., Desale, B. S., & Shinde, R. S. (2020). Efficient synthesis, spectroscopic and quantum chemical study of 2, 3-dihydrobenzofuran labelled two novel arylidene indanones: A comparative theoretical exploration. *Material Science Research India,* 17(2), 146-161.

[8] Adole, V. A. (2020). Synthetic approaches for the synthesis of dihydropyrimidinones/thiones (biginelli adducts): a concise review. *World journal of pharmaceutical research,* 9(6), 1067-1091.

[9] Adole, V. A., Koli, P. B., Shinde, R. A., & Shinde, R. S. (2020). Computational insights on molecular structure, electronic properties, and chemical reactivity of (E)-3-(4-chlorophenyl)-1-(2-hydroxyphenyl) prop-2-en-1-one. *Material Science Research India,* 17, 41-53.

[10] Adole, V. A., Jagdale, B. S., Pawar, T. B., & Desale, B. S. (2020). Molecular structure, frontier molecular orbitals, MESP and UV–visible spectroscopy studies of Ethyl 4-(3, 4-dimethoxyphenyl)-6-methyl-2-oxo-1, 2, 3, 4-tetrahydropyrimidine-5-carboxylate: A theoretical and experimental appraisal. *Material Science Research India,* 17(specialissue2020), 13-36.

[11] Shinde, R., & Adole, V. A. (2021). Anti-microbial evaluation, experimental and theoretical insights into molecular structure, electronic properties, and chemical reactivity of (E)-2-((1H-indol-3-yl) methylene)-2, 3-dihydro-1H-inden-1-one. *J Appl Organomet Chem,* 1(2), 48-58.

[12] Gong, H. H., Addla, D., Lv, J. S., & Zhou, C. H. (2016). Heterocyclic naphthalimides as new skeleton structure of compounds with increasingly expanding relational medicinal applications. *Current topics in medicinal chemistry,* 16(28), 3303-3364.

[13] Adole, V. A. (2021). Synthesis, antibacterial, antifungal and DFT studies on structural, electronic and chemical reactivity of (E)-7-((1H-Indol-3-yl) methylene)-1, 2, 6, 7-tetrahydro-8H-indeno [5, 4-b] furan-8-one. *Advanced Journal of Chemistry-Section A, 4*(3), 175-87.

[14] Shinde, R. A., Adole, V. A., Jagdale, B. S., & Pawar, T. B. (2021). Superfast synthesis, antibacterial and antifungal studies of halo-aryl and heterocyclic tagged 2, 3-dihydro-1 H-inden-1-one candidates. *Monatshefte für Chemie-Chemical Monthly,* 152(6), 649-658.

[15] Al-Harbi, N. O., Bahashwan, S. A., Fayed, A. A., Aboonq, M. S., & Amr, A. E. G. E. (2013). Anti-parkinsonism, hypoglycemic and anti-microbial activities of new poly fused ring heterocyclic candidates. *International journal of biological macromolecules, 57*, 165-173.

[16] Nerkar, A. U. (2021). Use of pyrimidine and its derivative in pharmaceuticals: A review. *Journal of Advanced Chemical Sciences,* 729-732.

[17] Sun, Y., Fu, R., Lin, S., Zhang, J., Ji, M., Zhang, Y., ... & Xu, H. (2021). Discovery of new thieno [2, 3-d] pyrimidine and thiazolo [5, 4-d] pyrimidine derivatives as

orally active phosphoinositide 3-kinase inhibitors. *Bioorganic & Medicinal Chemistry,* 29, 115890.

[18] Li, J., An, B., Song, X., Zhang, Q., Chen, C., Wei, S., ... & Zou, Y. (2021). Design, synthesis and biological evaluation of novel 2, 4-diaryl pyrimidine derivatives as selective EGFRL858R/T790M inhibitors. *European Journal of Medicinal Chemistry,* 212, 113019.

[19] Sun, J., Wan, Z., Chen, Y., Xu, J., Luo, Z., Parise, R. A., ... & Li, S. (2020). Triple drugs co-delivered by a small gemcitabine-based carrier for pancreatic cancer immunochemotherapy. *Acta biomaterialia,* 106, 289-300.

[20] Ochoa, S., & Milam, V. T. (2020). Modified nucleic acids: Expanding the capabilities of functional oligonucleotides. *Molecules,* 25(20), 4659.

[21] Kumar, B., Sharma, P., Gupta, V. P., Khullar, M., Singh, S., Dogra, N., & Kumar, V. (2018). Synthesis and biological evaluation of pyrimidine bridged combretastatin derivatives as potential anticancer agents and mechanistic studies. *Bioorganic chemistry,* 78, 130-140.

[22] Perupogu, N., Kumar, D. R., & Ramachandran, D. (2020). Anticancer activity of newly synthesized 1, 2, 4-oxadiazole linked 4-(oxazolo [5, 4-d]pyrimidine derivatives. *Chemical Data Collections,* 27, 100363.

[23] Al Safarjalani, O. N., Rais, R., Shi, J., Schinazi, R. F., Naguib, F. N., & el Kouni, M. H. (2006). Modulation of 5-fluorouracil host-toxicity and chemotherapeutic efficacy against human colon tumors by 5-(phenylthio) acyclouridine, a uridine phosphorylase inhibitor. *Cancer chemotherapy and pharmacology,* 58, 692-698.

[24] Lin, J., Roy, V., Wang, L., You, L., Agrofoglio, L. A., Deville-Bonne, D., ... & Eriksson, S. (2010). 3′-(1, 2, 3-Triazol-1-yl)-3′-deoxythymidine analogs as substrates for human and Ureaplasma parvum thymidine kinase for structure–activity investigations. *Bioorganic & medicinal chemistry,* 18(9), 3261-3269.

[25] Peters, G. J. (2014). Novel developments in the use of antimetabolites. Nucleosides, *Nucleotides and Nucleic Acids,* 33(4-6), 358-374.

[26] Garcia-Manero, G., Gore, S. D., Cogle, C., Ward, R., Shi, T., MacBeth, K. J., ... & Skikne, B. (2011). Phase I study of oral azacitidine in myelodysplastic syndromes, chronic myelomonocytic leukemia, and acute myeloid leukemia. *Journal of Clinical Oncology,* 29(18), 2521.

[27] Baughman, R. P., & Grutters, J. C. (2015). New treatment strategies for pulmonary sarcoidosis: antimetabolites, biological drugs, and other treatment approaches. *The Lancet Respiratory Medicine,* 3(10), 813-822.

[28] Hitchings, G. H., Elion, G. B., Vander Werff, H., & Falco, E. A. (1948). Pyrimidine derivatives as antagonists of pteroylglutamic acid. *Journal of Biological Chemistry,* 174, 765-766.

[29] Johns, D. G., Iannotti, A. T., Sartorelli, A. C., Booth, B. A., & Bertino, J. R. (1964). Enzymic oxidation of methotrexate and aminopterin. *Life Sciences,* 3(12), 1383-1388.

[30] Polat, M. F., & Tuncbilek, M. (2021). Highly efficient chemical phosphorylation of 6-(4-phenylpiperazine-1-yl)-9-(β-D-ribofuranosyl)-9 H-purine. *Nucleosides, Nucleotides & Nucleic Acids,* 40(3), 233-241.

[31] Tsukamoto, M., Yamashita, M., Nishi, T., & Nakagawa, H. (2019). A human ABC transporter ABCC4 gene SNP (rs11568658, 559 G> T, G187W) reduces ABCC4-dependent drug resistance. *Cells,* 8(1), 39.

[32] Basha, J., & Goudgaon, N. M. (2021). A comprehensive review on pyrimidine analogs-versatile scaffold with medicinal and biological potential. *Journal of Molecular Structure,* 1246, 131168.

[33] Gong, M., Yang, J., Li, Y., & Gu, J. (2020). Glutathione-responsive nanoscale MOFs for effective intracellular delivery of the anticancer drug 6-mercaptopurine. *Chemical Communications,* 56(47), 6448-6451.

[34] Li, H., An, X., Li, Q., Yu, H., & Li, Z. (2021). Construction and analysis of competing endogenous RNA network of MCF-7 breast cancer cells based on the inhibitory effect of 6-thioguanine on cell proliferation. *Oncology Letters,* 21(2), 1-1.

[35] Peters, G. J., Leyva, A., & Schwartsmann, G. (2020). Resistance to differentiation affects ribo-and deoxyribonucleotide pools and sensitivity to pyrimidine metabolism antagonists in HL60 cells. *Nucleosides, Nucleotides & Nucleic Acids,* 39(10-12), 1369-1378.

[36] Lin, E. S., & Huang, C. Y. (2021). Crystal structure of the single-stranded DNA-binding protein SsbB in complex with the anticancer drug 5-fluorouracil: Extension of the 5-fluorouracil interactome to include the oligonucleotide/oligosaccharide-binding fold protein. *Biochemical and Biophysical Research Communications,* 534, 41-46.

[37] Lamie, P. F., & Philoppes, J. N. (2020). 2-Thiopyrimidine/chalcone hybrids: design, synthesis, ADMET prediction, and anticancer evaluation as STAT3/STAT5a inhibitors. *Journal of enzyme inhibition and medicinal chemistry, 35*(1), 864-879.

[38] Yamane, M., & Toyooka, S. (2021). Role of surgery in a novel multimodal therapeutic approach to complete cure of advanced lung cancer: current and future perspectives. *Surgery today,* 1-11.

[39] Paroha, S., Verma, J., Dubey, R. D., Dewangan, R. P., Molugulu, N., Bapat, R. A., ... & Kesharwani, P. (2021). Recent advances and prospects in gemcitabine drug delivery systems. *International Journal of Pharmaceutics,* 592, 120043.

[40] Zacharski, L. R., Moritz, T. E., Baczek, L. A., Rickles, F. R., Edwards, R. L., Forman, W. B., ... & Hoppel, C. L. (1988). Effect of mopidamol on survival in carcinoma of the lung and colon: final report of Veterans Administration Cooperative Study No. 188. *JNCI: Journal of the National Cancer Institute,* 80(2), 90-97.

[41] Kang, Y., Saito, M., & Toyoda, H. (2020). Molecular and regulatory mechanisms of desensitization and resensitization of GABAA receptors with a special reference to propofol/barbiturate. *International Journal of Molecular Sciences,* 21(2), 563.

[42] Hamedi, S., Forouzanfar, F., Rakhshandeh, H., & Arian, A. (2019). Hypnotic effect of Portulaca oleracea on pentobarbital-induced sleep in mice. *Current Drug Discovery Technologies,* 16(2), 198-203.

[43] Wierzba, W., Müller, D., Richter, K., & Klinger, W. (1984). The influence of cyclobarbital and diazepam on drug metabolism in vitro and their binding to cytochrome P-450. *Biomedica Biochimica Acta,* 43(12), 1425-1430.

[44] Mamina, E. A., & Bolotov, V. V. (2004). Analysis of barbituric acid derivatives in biological objects. *Pharmaceutical Chemistry Journal,* 38(10), 578-581.
[45] Pacifici, G. M. (2016). Clinical pharmacology of phenobarbital in neonates: effects, metabolism and pharmacokinetics. *Current pediatric reviews,* 12(1), 48-54.
[46] Sun, J. G., Ruan, F., Zeng, X. L., Xiang, J., Li, X., Wu, P., ... & Liu, F. Y. (2016). Clitocine potentiates TRAIL-mediated apoptosis in human colon cancer cells by promoting Mcl-1 degradation. *Apoptosis,* 21, 1144-1157.
[47] Larwood, D. J. (2020). Nikkomycin Z—ready to meet the promise?. *Journal of Fungi,* 6(4), 261.
[48] Zilles, J. L., Croal, L. R., & Downs, D. M. (2000). Action of the thiamine antagonist bacimethrin on thiamine biosynthesis. *Journal of bacteriology,* 182(19), 5606-5610.
[49] Leviev, I. G., Rodriguez-Fonseca, C., Phan, H., Garrett, R. A., Heilek, G., Noller, H. F., & Mankin, A. S. (1994). A conserved secondary structural motif in 23S rRNA defines the site of interaction of amicetin, a universal inhibitor of peptide bond formation. *The EMBO Journal,* 13(7), 1682-1686.
[50] Bu, Y. Y., Yamazaki, H., Ukai, K., & Namikoshi, M. (2014). Anti-mycobacterial nucleoside antibiotics from a marine-derived Streptomyces sp. TPU1236A. *Marine Drugs,* 12(12), 6102-6112.
[51] Kollatos, N., Mitsos, C., Manta, S., Tzioumaki, N., Giannakas, C., Alexouli, T., ... & Komiotis, D. (2020). Design, synthesis, and biological evaluation of novel C5-modified pyrimidine ribofuranonucleosides as potential antitumor or/and antiviral agents. *Medicinal Chemistry,* 16(3), 368-384.
[52] Acosta, P., Insuasty, B., Ortiz, A., Abonia, R., Sortino, M., Zacchino, S. A., & Quiroga, J. (2016). Solvent-free microwave-assisted synthesis of novel pyrazolo [4′, 3′: 5, 6]pyrido [2, 3-d]pyrimidines with potential antifungal activity. *Arabian Journal of Chemistry,* 9(3), 481-492.
[53] Krečmerová, M., Dračínský, M., Snoeck, R., Balzarini, J., Pomeisl, K., & Andrei, G. (2017). New prodrugs of two pyrimidine acyclic nucleoside phosphonates: Synthesis and antiviral activity. *Bioorganic & Medicinal Chemistry,* 25(17), 4637-4648.
[54] Topalis, D., Gillemot, S., Snoeck, R., & Andrei, G. (2018). Thymidine kinase and protein kinase in drug-resistant herpesviruses: Heads of a Lernaean Hydra. *Drug Resistance Updates,* 37, 1-16.
[55] Vashishtha, A. K., & Kuchta, R. D. (2015). Polymerase and exonuclease activities in herpes simplex virus type 1 DNA polymerase are not highly coordinated. *Biochemistry,* 54(2), 240-249.
[56] Fuchs, W., Fichtner, D., Bergmann, S. M., & Mettenleiter, T. C. (2011). Generation and characterization of koi herpesvirus recombinants lacking viral enzymes of nucleotide metabolism. *Archives of virology,* 156, 1059-1063.
[57] Ayisi, N. K., Gupta, V. S., Meldrum, J. B., Taneja, A. K., & Babiuk, L. A. (1980). Combination chemotherapy: interaction of 5-methoxymethyldeoxyuridine with adenine arabinoside, 5-ethyldeoxyuridine, 5-iododeoxyuridine, and phosphonoacetic acid against herpes simplex virus types 1 and 2. *Antimicrobial Agents and Chemotherapy,* 17(4), 558-566.

[58] Kurmi, M., Sahu, A., Balhara, A., Singh, I. P., Kulkarni, S., Singh, N. K., ... & Singh, S. (2020). Stability behaviour of antiretroviral drugs and their combinations. 11: Characterization of interaction products of zidovudine and efavirenz, and evaluation of their anti HIV-1 activity, and physiochemical and ADMET properties. *Journal of Pharmaceutical and Biomedical Analysis,* 178, 112911.

[59] Ho, S., Wong, J. G., Ng, O. T., Lee, C. C., Leo, Y. S., Lye, D. C. B., & Wong, C. S. (2020). Efficacy and safety of abacavir/lamivudine plus rilpivirine as a first-line regimen in treatment-naïve HIV-1 infected adults. *AIDS Research and Therapy,* 17(1), 1-9.

[60] Pinching, A. J. (1991). HIV/AIDS pathogenesis and treatment: new twists and turns. *Current Opinion in Immunology,* 3(4), 537-542.

[61] Harrison, C. (2020). Coronavirus puts drug repurposing on the fast track. *Nat Biotechnol,* 379-381.

[62] Zequn, Z., Yujia, W., Dingding, Q., & Jiangfang, L. (2021). Off-label use of chloroquine, hydroxychloroquine, azithromycin and lopinavir/ritonavir in COVID-19 risks prolonging the QT interval by targeting the hERG channel. *European Journal of Pharmacology,* 893, 173813.

[63] Kandil, S., Pannecouque, C., Chapman, F. M., Westwell, A. D., & McGuigan, C. (2019). Polyfluoroaromatic stavudine (d4T) ProTides exhibit enhanced anti-HIV activity. *Bioorganic & Medicinal Chemistry Letters,* 29(24), 126721.

# Chapter 3

# The Evolution of Pyrimidine Bases

**Özgür Kebabci**[1,*]
**and Burak Tüzün**[2,†]
[1]Department of Molecular Biology and Genetics, Science Faculty, Sivas Cumhuriyet University, Sivas, Turkey
[2]Plant and Animal Production Department, Technical Sciences Vocational School of Sivas, Sivas Cumhuriyet University, Sivas, Turkey

## Abstract

Evolution occurs when genes in a population of cells change in sequence and frequency over time and dominate by modification. However, in the evolutionary process, before the formation of cells, there must also be compounds that form these cells. One of these compounds should be pyrimidine bases. Pyrimidine bases are the bases that must be present in every cell, in the structure of DNA and RNA. In this case, the basic question to be asked is how did the pyrimidine bases form on the early Earth, in other words, how did they evolve? Many compounds may have formed on the Early Earth, and some of these compounds formed the origin of life. There is strong evidence that RNA evolved first, and then DNA. Both of these contain purine and pyrimidine bases as building blocks. It is widely accepted that purines and pyrimidines, the building blocks that gave rise to life on our planet, formed on early abiotic Earth through multicomponent reactions (MCRs). Lasty, all molecules found in the reviewer will be examined for their activities against Crystal Structure of the human nucleosome containing the H2B E76K mutant

* Corresponding Author's E-mail: ozgur.kebabci@gmail.com. ORCID ID: 0000-0002-9404-747X.
† Corresponding Author's E-mail: theburaktuzun@yahoo.com. http://orcid.org/0000-0002-0420-2043.

In: The Chemistry of Pyrimidine Derivatives
Editor: Barry Schneider
ISBN: 979-8-89113-563-5

protein (PDB ID: 5Y0D) and structure of a B-DNA dodecamer protein (PDB ID: 1BNA).

**Keywords:** evolution, pyrimidine, abiotic Earth, MCRs

## 1. Introduction

Millions and even billions of species, along with those that we have not yet identified, live on the planet we call Earth. In addition to simple single-celled life forms such as microorganisms, multicellular and complex-structured living organisms are also home to this planet. All life forms have adapted to a symbiosis together on planet Earth, leaving aside human beings. The first question should be asked is "What is the source of life?". Is the origin of life in the precursors of macromolecules? A macromolecule is a very large molecule consisting of thousands of covalently bonded atoms, important for biophysical processes such as a protein or nucleic acid, and many macromolecules are polymers of smaller molecules called monomers. The most common macromolecules are biopolymers such as nucleic acids, proteins and carbohydrates. In addition, lipids are also large molecules that are not polymeric (Berg et. al. 2019). The source of life, on the other hand, should be in the precursors of all these macromolecules, without distinction from each other. Another question should be asked is, "How did these macromolecules form amino acids, nucleotides and simple sugars, which are the precursors of proteins, nucleic acids (DNA and RNA) and polysaccharides, respectively?". The answer to this question is hidden in early Earth conditions. The answer to this question is hidden in the early world conditions. In terms of nucleotides, which are organic molecules consisting of a nitrogenous base (called a nucleobase), a pentose sugar, and a phosphate, it is difficult to say that these are also chemically simple molecules. When we take the situation further, although the basic skeletal structures are similar in nucleobases, there are different structures such as guanine, adenine (purines), cytosine and thymine (pyrimidines) in DNA, and uracil instead of thymine in RNA. Even in this case, the presence of uracil pyrimidine instead of thymine pyrimidine in RNA shows the difference more clearly. All these structures are indispensable molecules of today's organisms. However, how the formation of these molecules took place in early Earth conditions is still a mystery, although it is supported by various hypotheses.

## 1.1. A Brief View At The Early World

Earth was formed about 4.5 billion years ago. Our knowledge and assumptions about the Early Earth are based on analyzes of chemical and slowly decaying radioactive isotopes of ancient rocks and minerals. At first glance, planet Earth should look like a planet consisting of hot magma. As the temperature decreased over time, a metallic core, a rocky mantle, and a thin surface crust of lower density were formed. Early Earth is quite different from today's Earth and is characterized by a molten surface under intense bombardment by asteroids and other objects from space. These conditions, in which life was not possible, probably lasted for 500 million years. The fact that there was intense heat during this period, and moreover, the absence of liquid water, shows that the Earth was devoid of any life forms.

Certainly there was no oxygen on the early Earth either. Therefore, an anoxic environment prevails. Just as there is no oxygen, the existence of the ozone layer cannot be mentioned. Perhaps even more important than oxygen is the lack of liquid water on the early Earth. It is already known that Cyanobacteria are responsible for the atmospheric level of oxygen on Earth. But the necessity of liquid water for life is absolute. In anaerobic periods, life already exists in anoxic Earth conditions, and liquid water must have been present. It is thought that the water on Earth is caused by the volcanic gas outflow of the planet's interior and numerous collisions with icy comets and asteroids. Of course, it should be noted that this water is also present as water vapor when we consider the temperature of the Earth.

No rocks belonging to the early periods that will shed light on the origin of the Earth have yet been discovered. The probable reason for this is that these rocks have undergone geological metamorphosis. However, the ancient crystals of the mineral zircon (ZrSiO4), which were discovered to have been formed in the early times of the Earth, have provided insight into the conditions in the periods before the formation of life on Earth. As a result of the analysis of these ancient zircon crystals, it has been stated that a solid crust and liquid water existed on Earth 4.3 billion years ago. Another evidence of the existence of water is sedimentary rocks, because sedimentary rocks are formed only under water. The oldest sedimentary rocks that have survived to the present day have been discovered in southwestern Greenland and are 3.86 billion years old. These rocks, which are evidence for the existence of early oceans, contain fossilized cell remains and also carbon isotope ratios, which provide evidence for life. This and many similar evidences show that liquid water appeared 4 billion years ago, and it is known that the first life forms appeared shortly after that.

## 1.2. Formation of Pyrimidines And Some Hypotheses Related To Pyrimidines

The origin of life on Earth remains the greatest of mysteries hidden by the depths of time, as very few rocks have survived to bear witness to this period of Earth's history. Experimental evidence suggests that organic molecules, including RNA nucleotides, amino acids, and lipids, could have formed spontaneously in early Earth conditions, providing the necessary preconditions for the first living systems. However, the abiotic formation of these organic molecules necessary for the first living systems seems unlikely due to the conditions on the Earth's surface. Because the Earth's surface has extremely high temperatures and high levels of ultraviolet radiation, the formation of organic molecules, or more accurately, life, is unlikely to occur under these conditions.

In principle, the origin of life can be studied upstream from geochemistry or downstream from biology, but in practice there are problems with both approaches. A large amount of chemistry is potentially possible in a submarine vent or a drying lagoon or an impact crater or a reduced atmosphere exposed to lightning or any other scenario imaginable (Sutherland, 2016). In the past, scientists have suggested that one or the other subsystem existed first, followed by the others, which explains the multiplicity of existing hypotheses. For example, "RNA must have been formed first" (Joyce, 2002), "Life cannot be possible without building blocks and energy, therefore metabolism must have existed first" (Wächtershäuser, 1992), "Proteins must have come first because genetics and metabolism cannot occur without catalysis" (Plankensteiner et. al., 2005) or "Membranes must have existed in the beginning because it is difficult to imagine the evolution of Darwin selection without compartments" (Segré et.al., 2001). All these views have in themselves logical and supporting evidence.

If the self-formation of organic molecules for primitive life did not take place on the Earth's surface, where should it have taken place? One hypothesis that has received the most support clarifies this situation. According to this hypothesis, it is suggested that life was formed in hydrothermal vents located on the ocean floor (Figure 1). It is more likely to be shielded from deadly UV radiation in the depths of the ocean than on the Earth's surface. In addition, according to the hypothesis, it is suggested that in these hydrothermal systems, continuous and abundant energy takes place in the form of reduced inorganic compounds such as Hydrogen ($H_2$), hydrogen sulfide ($H_2S$) and elemental sulfur ($S^0$). The geochemistry of these hydrothermal vents supports the abiotic production of molecules such as amino acids, nucleotides, sugars and lipids that are critical to the emergence of life. Also, although there is no biological membrane structure yet to conserve energy, parts in these systems may have provided it. Whether on the seafloor or elsewhere, some kind of prebiotic chemistry must have facilitated the development of the first self-replicating systems, the precursors of cellular life (Madigan et. al., 2021).

It has been suggested in the Oparin-Haldane hypothesis that chemical evolution took place for a long time before the emergence of life (Oparin, 1924; Haldane, 1929; Ponnamperuma & Gabel, 1968; Tirard, 2017). In the formation of life, some biomolecules are at the forefront compared to others. There are two important factors in terms of chemical evolution on the primitive Earth. One of these is self-replicating systems. The other one is the accumulation of organic molecules. One of the self-replicating systems is RNA, while the other is DNA. Today, DNA is at the forefront, but what about in the early world, when cellular life was not yet in question? According to one hypothesis, which is the "RNA World" hypothesis, it is more likely that life started in an RNA world instead of DNA (Figure 2). RNA molecules were probably a central component of the first self-replicating systems. RNA molecules can perform a variety of biochemical functions, as well as to bind small molecules such as nucleotides and amino acids, to catalyze a number of simple biochemical reactions, and to serve as templates for their own synthesis. In addition, RNA can also catalyze protein synthesis. Therefore, it is not unreasonable to imagine that the earliest forms of life were based solely on RNAs and had little or no need for DNA or proteins. However, this situation should not have lasted too long, as today, proteins undertake the catalytic reactions and DNA, which is more stable than RNA by nature, undertakes the transfer of genetic information. DNA, on the other hand, should become a template for RNA synthesis.

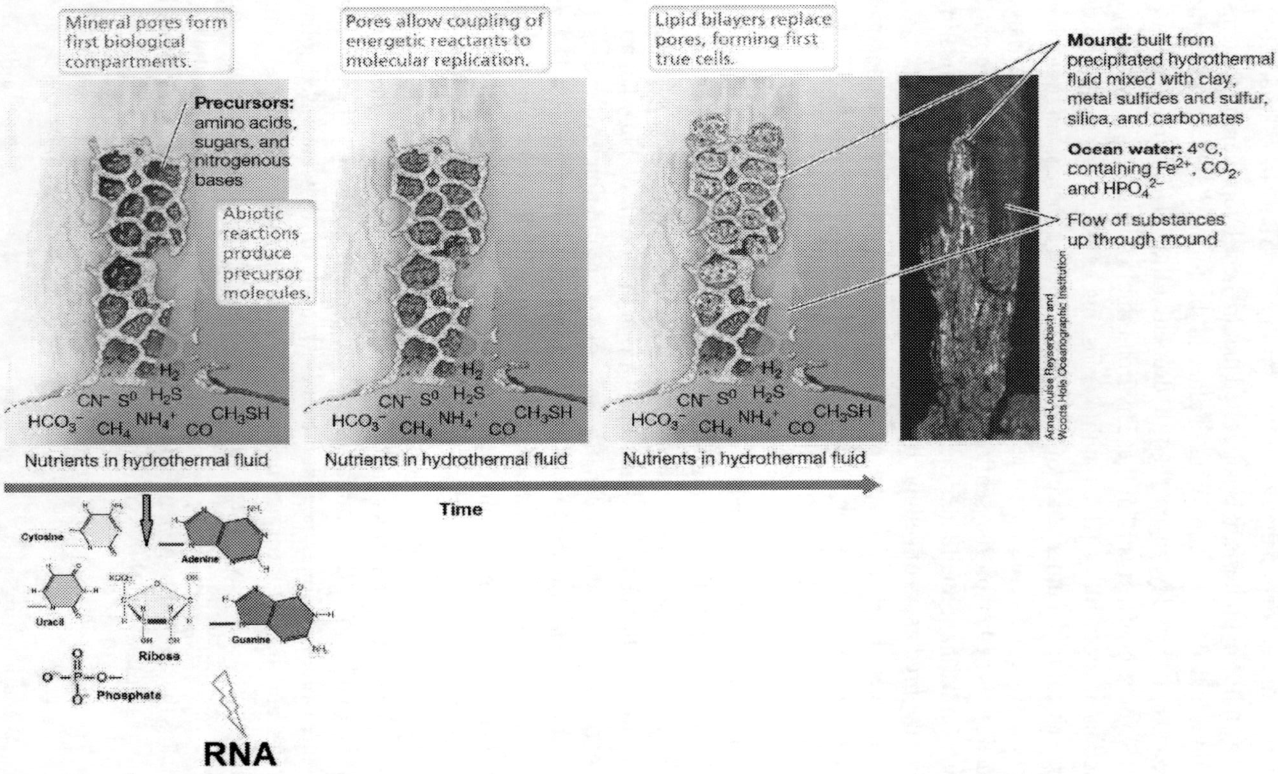

Mineral pores form first biological compartments.
Pores allow coupling of energetic reactants to molecular replication.
Lipid bilayers replace pores, forming first true cells.
Precursors: amino acids, sugars, and nitrogenous bases
Abiotic reactions produce precursor molecules.
Mound: built from precipitated hydrothermal fluid mixed with clay, metal sulfides and sulfur, silica, and carbonates
Ocean water: 4°C, containing $Fe^{2+}$, $CO_2$, and $HPO_4^{2-}$
Flow of substances up through mound
Anna-Louise Reysenbach and Woods Hole Oceanographic Institution
$H_2$ $CN^-$ $S^0$ $H_2S$ $HCO_3^-$ $CH_4$ $NH_4^+$ $CO$ $CH_3SH$
Nutrients in hydrothermal fluid
Nutrients in hydrothermal fluid
Nutrients in hydrothermal fluid
Time
Cytosine
Adenine
Uracil
Ribose
Guanine
Phosphate
RNA


**Figure 1.** (Continued).

**Figure 1.** The hydrothermal vents on the ocean floor and formations of the origin of life. Deep-sea hydrothermal vents may have created conditions necessary for Earth's life. In addition to being stable environments suitable for life, these vents contain a source of chemical energy, precursors for the formation of biomolecules and compartments for biochemical reactions that take place before cell formation, and mineral pores. (a) Abiotically formed precursor molecules accumulate in mineral pores. (b) Chemical gradients occur in mineral pores. This provides an energy source necessary for the proliferation of precellular biomolecules. (c) Subsequently formed membranes replace the mineral partitions and lead to the formation of the first cells. (d) Photograph of a hydrothermal vent located today. The mineral-rich, warm hydrothermal fluid mixes with cold ocean water and eventually forms precipitates composed of Fe and S compounds, clays, silicates, and carbonates. These mineral deposits form pores that can serve as energy-rich compartments to facilitate the evolution of precellular life forms. The presence of pyrimidines in the mireal pores located in these hydrothermal vents has already been demonstrated. (Modified from Brock, Biology of Organisms, Madigan et. al., 2021).

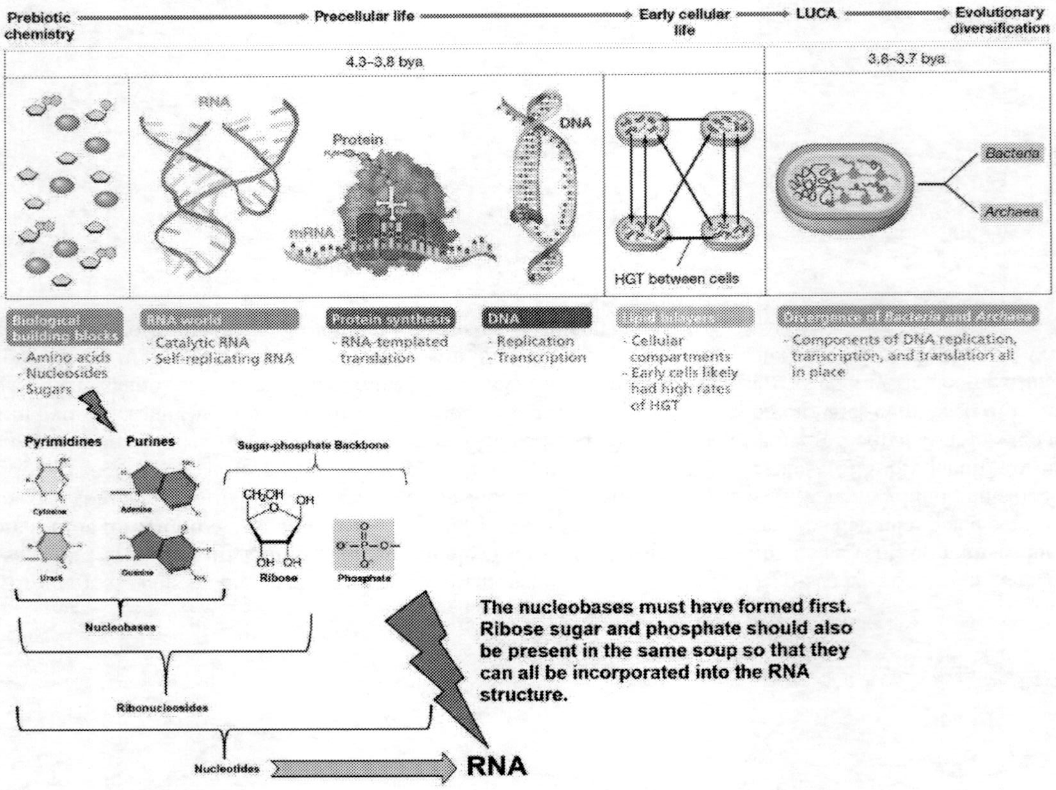
Prebiotic chemistry
Precellular life
Early cellular life
LUCA
Evolutionary diversification
4.3–3.8 bya
3.8–3.7 bya
RNA
Protein
mRNA
DNA
HGT between cells
Bacteria
Archaea
Biological building blocks
- Amino acids
- Nucleosides
- Sugars
RNA world
- Catalytic RNA
- Self-replicating RNA
Protein synthesis
- RNA-templated translation
DNA
- Replication
- Transcription
- Cellular compartments
- Early cells likely had high rates of HGT
Divergence of Bacteria and Archaea
- Components of DNA replication, transcription, and translation all in place
Pyrimidines
Purines
Sugar-phosphate Backbone
Ribose
Phosphate
Nucleobases
Ribonucleosides
Nucleotides
RNA
The nucleobases must have formed first. Ribose sugar and phosphate should also be present in the same soup so that they can all be incorporated into the RNA structure.

**Figure 2.** (Continued).

**Figure 2.** Events that are assumed to have occurred before cellular life. The formation of the first cell, LUCA (Last Universal Common Ancestor), dates back an estimated 3 7-3.8 billion years. Before the formation of cellular forms, about 500 million years ago (4.3 billion years ago), the first biological building blocks must have formed. The earliest self-replicating biological systems can then be based on catalytic RNA. At some point in this process, RNA enzymes may have developed the ability to synthesize proteins, and these proteins may have become major catalytic molecules. The formation of the DNA-based genome requires the evolution of DNA and RNA polymerases. Subsequently, a lipid bilayer must have formed in early cellular life. The formation of this structure is important in two respects. First is the electron transport region, which is important for energy conservation. The other is the inclusion and protection of biomolecules. LUCA (last universal common ancestor) has a lipid bilayer and is a cellular organism capable of using DNA, RNA, and proteins. It probably took millions of years after LUCA and before the divergence of Bacteria and Archaea. (Modified from Brock, Biology of Organisms, Madigan et. al. 2021).

This "RNA World" hypothesis is perhaps one of the most accepted hypotheses and points to the existence of an RNA world that existed on Earth before modern cells arose (Nam et.al, 2018). The RNA world hypothesis postulates that RNA plays the role of DNA (genotype) and proteins (phenotype) and then gives rise to an RNA/DNA/Protein world, implying that proteins and DNA are inventions of an RNA world (Gavette et. al., 2016). The RNA world hypothesis predicts that life began with RNAs that can recognize (self) and reproduce (Becker et. al., 2019). To understand the origin of RNA, it is necessary to consider which molecules may have originally acted instead of RNA substructures (i.e., nucleobases (recognition units), ribose (a three-functional binder), and phosphate (an ionized binder) in the earliest ancestor of RNA (or proto-RNA) (Cafferty & Hud, 2015). The missing link of the RNA hypothesis is how ribonucleosides, the building blocks of RNA, evolved from simpler molecules. It is widely accepted that purines and pyrimidines, the building blocks that gave rise to life on our planet, formed through multicomponent reactions (MCRs) on early abiotic Earth. Purine and pyrimidine are the most widely distributed N heterocycles in nature and play a central role in terrestrial biology and origin of life (Martina et. al., 2023). Purine and pyrimidine nitrogen bases have been synthesized in vitro by many scientists using prebiotic compounds (Zubay & Mui, 2001; Hill & Orgel, 2002). Ribose has also been synthesized in the same way (Ricardo et al., 2004; Saad, 2018). All these studies are aimed at supporting the RNA World hypothesis. The prerequisite for the RNA world is the formation of RNA under prebiotic conditions, and the first step requires the simultaneous formation of pyrimidine and purine nucleosides in the existing environment.

The origin of pyrimidine and purine nucleobases in a prebiotic environment can be traced back to several key starting molecules. In the 60s, many studies were conducted on the synthesis of pyrimidine. in 1961, Fox and Harada heated malic acid and urea and eventually obtained uracil (Fox & Harada, 1961; Ponnamperuma & Gabel, 1968). A number of pyrimidine syntheses have been established using cyanoacetylene. Cytosine was obtained by fusing cyanoacetylene with urea. Similarly, uracil was prepared from propiolamide. The chemistry leading to the formation of pyrimidines begins with cyanoacetylene as the basic building block, which has also been detected in the atmosphere of interstellar Titan (Thaddeus, 2006). It has been suggested that molecules such as cyanoacetylene are prebiotic starting materials that may have formed at significant concentrations in hydrothermal vents (Rimmer, 2019). Apart from cyanoacetylene, salts (nitrides, carbonates and borates) and metals (Zn or Fe) such as $NH_3$, $NH_2OH_2$, hydroxylurea, (hydroxyamino) methanedisulfonate, HCN, urea, formic acid and isocyanate have also been reported to be the origin of pyrimidine and purine nucleobases (Becker et. al. ., 2019). Suppose that purine and pyrimidine bases were formed under early Earth conditions. After that, the ribonucleosides and nucleotides required for RNA must also have been formed in some way. For ribonucleosides, ribose sugar must already exist in the environment. For nucleotides, a phosphate group must be added to this structure, which must also have been formed in the environment (Figure 2). It has been suggested that Earth's primitive reducing atmosphere is at least somewhat transparent between 2400 Å and 2900 Å, and that photoactivation of purines and pyrimidines, which absorb ultraviolet light, is a possible step in the formation of nucleosides and nucleotides. (Sagan, 1961).

However, there are other hypotheses other than the RNA World hypothesis. Scientists who argue that RNA did not come first have suggested that proteins or metabolism or even membranes came first. Although some scientists suggest that all this may have occurred at the same time, this does not seem statistically possible. There are four topics that are subject to the objections of critics of the RNA hypothesis. The first of these is that RNA is too complex a molecule to occur prebiotically. Second, RNA is an inherently unstable molecule. The third is due to the catalysis feature of RNA, a feature that is rarely seen and only exists in long RNA sequences. Finally, another objection is that the catalytic repertoire of RNA is rather limited (Bernhardt, 2012). We will not discuss all these different hypotheses here, as our topic is pyrimidines. However, it is helpful to mention a hypothesis that also includes pyrimidines.

This hypothesis is that organic compounds, including pyrimidines, have been detected in various meteorites and comets, and that these organic compounds came to Earth as a result of the impact of meteorites and parts from comets. For instance, Arrhenius suggested that panspermia, which exists scattered in the cosmic space, sprouts life if the conditions are suitable as a result of the falling of life-bearing seeds onto the planets (Horowitz et. al., 1962). Oró (1963), theorized that pyrimidine formation could occur from C-3 molecular species found in comets. Carbon suboxide ($C_3O_2$) is a compound detected in cosmic bodies and may be a suitable route for pyrimidine synthesis (Oró, 1961). Many celestial bodies and comets have been investigated for the existence of organic molecules or their building blocks. Many nitrogen heterocycles have been detected in carbonaceous meteorites. Among the nitrogen heterocycles detected are pyridine carboxylic acids, diketopiperazine, hydantoins, purines, pyrimidines, triazines, pyridines, quinolines, carboxylactams, lactams, lactims, and the amino acid proline. Murchison meteorite is also a meteorite that has been the subject of research. Although cytosine could not be detected in Murchison meteorite, adenine, uracil and guanine, xanthine and hypoxanthine were detected (Levy & Miller, 1998). The abundance of extraterrestrial amino acids found in the Murchison and the low level of terrestrial amino acid contaminants support the idea that purines and pyrimidines are unique to the meteorite (Glavin & Bada. 2004). The efficiency of the extraction procedure for purines, pyrimidines and triazines was determined by extracting a sample from the Allende meteorite (CV3) with known amounts of standard compounds added. A conclusion about the origin of purines and pyrimidines detected in meteorites was reached in 2008 when compound-specific carbon isotope measurements of these compounds were made using gas chromatography-combustion-isotope ratio mass spectrometry (GC-C-IRMS) (Martins, 2018). Pyrimidines have also been detected in many other meteorites such as Y-74662 (CM2) and Y-791198 (CM2). Some organic molecules have been detected in the tail dust of Halley's Comet by the Giotto spacecraft mass spectrometer. Pyridine and pyrimidine are among these organic molecules (Oró, 2004).

## 1.3. The Structure and the Discovery of Pyrimidines

Diazines are a group of organic compounds and have the molecular formula $C_4H_4N_2$. In diazines that have a benzene ring, two of the C-H moieties have been replaced by isolobal nitrogen. There are 3 structural isomers of diazines: pyridazine with nitrogen atoms at positions 1 and 2, pyrazine at positions 1 and 4, and finally pyrimidine at positions 1 and 3 (Figure 3). Pyrimidines, whose IUPAC designation is 1,3-Diazabenzene, is an aromatic, heterocyclic, organic compound. They are an integral part of DNA and RNA. While the pyrimidines in the RNA structure are cytosine and uracil, the thymine pyrimidine is located instead of uracil in the DNA structure. The variety of compounds containing pyrimidine rings is substantial. Pyrimidine and its derivatives have many biological activities such as anticancer, antiviral, antimicrobial, anti-inflammatory, analgesic, antioxidant and antimalarial (Kumar & Narasimhan, 2018). Purines are also present in DNA and RNA. Adenine and guanine are located in both of them. In DNA, adenine pairs with thymine, while in RNA adenine pairs with uracil. In both DNA and RNA, guanine pairs with cytosine. Like pyrimidines purines are heterocyclic aromatic organic compounds. However, purines have an imidazole ring fused to the pyrimidine ring (Alberts et. al., 2002). The pyrimidine ring in purines is quite similar to those in pyrimidines. Could it be possible that first pyrimidines and then purines were formed on the early Earth in chemical evolution (Figure 4)?

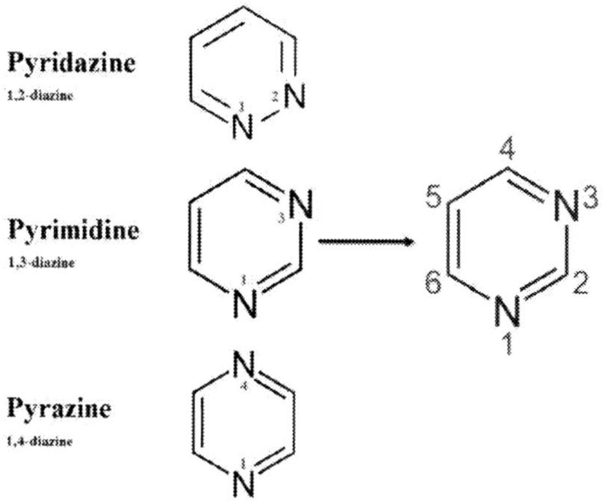
Pyridazine
1,2-diazine
Pyrimidine
1,3-diazine
Pyrazine
1,4-diazine

**Figure 3.** Diazines.

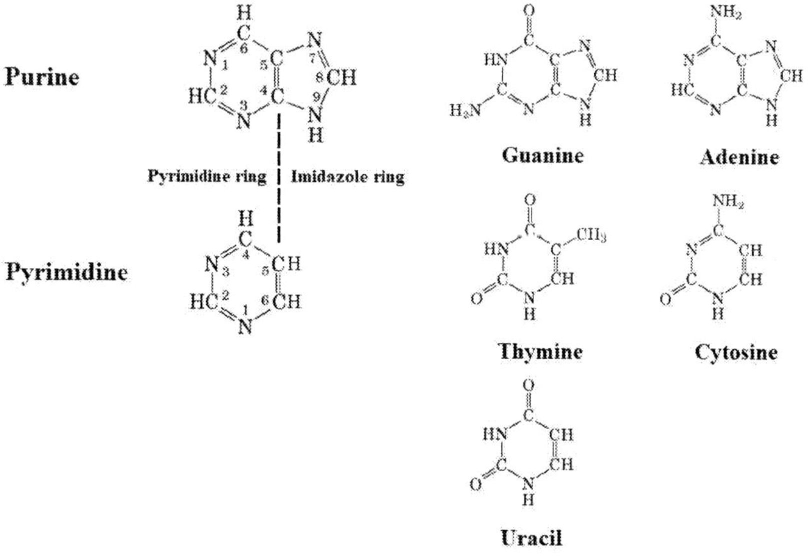
Purine
Pyrimidine
Pyrimidine ring
Imidazole ring
Guanine
Adenine
Thymine
Cytosine
Uracil

**Figure 4.** The structure of purine and pyrimidine rings and the similarity of purine and pyrimidine bases.

Although it is stated that the synthesis of pyrimidine derivatives started with Pinner's synthesis of pyrimidine derivatives by condensation of ethyl acetoacetate with amidines in 1884, Brugnatelli isolated the first pyrimidine derivative, which was named alloxan after the oxidative degradation of uric acid with nitric acid (Brown et. al. 1994). The story of synthesizing pyrimidine derivatives that began in 1818 continued in 1848 with Frankland and Kolbe pioneering the synthesis of a second pyrimidine. In their experiment, they obtained 2,6-diethyl-5-methyl-4-pyrimidinamine by heating propionitrile with metallic potassium. The third pyrimidine synthesis was carried out by Grimaux 30 years later. In 1887, barbituric acid was obtained from urea and malonic acid in the presence of phosphorus oxychloride. Pinner had the privilege of accepting the true nature of pyrimidine as a hexagonal analog of benzene and pyridine, as a result of his research in 1884, and first proposed the name "pyrimidine" in 1885. Widmann's suggestion of the name "miazin(e)" in 1888, perhaps more logically, was not widely accepted. In 1899, the parent compound was obtained by Gabriel and Colman. In their research, they succeeded in synthesizing the parent compound by converting barbituric acid to 2,4,6-trichloropyrimidine and then reducing it in hot water using zinc powder. An international pyrimidine research period was resumed in 1935. In the following years, studies continued in this direction due to the determination of the pharmacological and agrochemical activities of pyrimidines. Today Studies on pyrimidine derivatives especially pharmacologically with anticancer activity continue uninterruptedly.

## 1.4. Theoretical Calculations

An important method used to determine the molecules with the highest activity against biological materials is molecular docking. molecular docking calculations are made in Schrödinger's Maestro Molecular modeling platform (version 12.8) (Schrödinger Release 2021-3a). In the calculations made with this method, it is possible to comment on the active sites of the molecules. Calculations are made up of several steps. It first uses the protein preparation module (Schrödinger Release 2021-3b) to prepare the protein, then the LigPrep module (Schrödinger Release 2021-3c) to prepare the molecules. In order to interact the prepared proteins and molecules, they interact with each other with the Glide ligand docking tool (Shahzadi et al., 2022). Finally, the Qik-prop module of the Schrödinger software (Schrödinger Release 2021-3d) was used while performing ADME/T analysis (absorption, distribution, metabolism, excretion, and toxicity) in order to examine the effects and effects of the studied molecules on human metabolism.

The activities of the above molecules against Crystal Structure of the human nucleosome containing the H2B E76K mutant protein (PDB ID: 5Y0D) (Arimura et al., 2018) and structure of a B-DNA dodecamer protein (PDB ID: 1BNA) (Drew et al., 1981) were investigated. As a result of the calculations made in Table 1, the theoretical activities of these molecules were examined by molecular docking calculations, which calculations were made both to predict the activities of the molecules and to find the active sites of the molecules in the Figure 5-8 (Günsel et al., 2019; Günsel et al., 2020; Türkan et al., 2021).

**Table 1. Numerical values of the docking parameters of molecule against enzymes**

| 1BNA | Docking Score | Glide ligand efficiency | Glide hbond | Glide evdw | Glide ecoul | Glide emodel | Glide energy | Glide einternal | Glide posenum |
|---|---|---|---|---|---|---|---|---|---|
| 1 | -4.14 | -0.69 | 0.00 | -10.80 | -2.80 | -16.99 | -13.60 | 0 | 196 |
| 2 | -3.30 | -0.55 | -0.11 | -8.00 | -1.74 | -11.95 | -9.74 | 0 | 98 |
| 3 | -2.95 | -0.49 | 0.00 | -8.21 | -1.86 | -12.01 | -10.07 | 0 | 10 |
| 4 | -3.72 | -0.41 | -0.32 | -11.68 | -4.62 | -20.47 | -16.31 | 0 | 275 |
| 5 | -4.68 | -0.43 | -0.46 | -12.75 | -7.91 | -27.16 | -20.66 | 0 | 35 |
| 6 | -4.06 | -0.41 | -0.27 | -10.34 | -5.96 | -20.90 | -16.30 | 0 | 125 |
| 7 | -4.99 | -0.55 | -0.32 | -17.00 | -3.42 | -26.50 | -20.41 | 0 | 211 |
| 8 | -4.14 | -0.52 | -0.17 | -8.31 | -7.62 | -20.45 | -15.93 | 0 | 54 |
| 9 | -4.14 | -0.52 | -0.33 | -10.88 | -6.61 | -21.93 | -17.49 | 0 | 266 |
| 5YOD | Docking Score | Glide ligand efficiency | Glide hbond | Glide evdw | Glide ecoul | Glide emodel | Glide energy | Glide einternal | Glide posenum |
| 1 | -4.00 | -0.67 | 0.00 | -12.52 | -1.00 | -16.80 | -13.52 | 0 | 231 |
| 2 | -4.28 | -0.71 | 0.00 | -13.16 | -1.69 | -18.72 | -14.85 | 0 | 324 |
| 3 | -4.24 | -0.71 | 0.00 | -15.18 | -0.98 | -20.17 | -16.16 | 0 | 74 |
| 4 | -5.72 | -0.64 | -0.28 | -14.22 | -4.31 | -26.59 | -18.53 | 0 | 295 |
| 5 | -5.32 | -0.48 | -0.07 | -21.68 | -3.34 | -33.38 | -25.02 | 0 | 153 |
| 6 | -4.53 | -0.45 | 0.00 | -21.02 | 0.09 | -26.87 | -20.93 | 0 | 352 |
| 7 | -6.11 | -0.68 | -0.48 | -14.34 | -6.31 | -28.35 | -20.65 | 0 | 313 |
| 8 | -5.10 | -0.64 | -0.29 | -18.34 | -3.04 | -28.03 | -21.38 | 0 | 303 |
| 9 | -4.93 | -0.62 | -0.20 | -20.13 | -2.32 | -29.36 | -22.46 | 0 | 206 |

In molecular docking calculations, pyridazine (1), pyrimidine (2), pyrazine (3), purine (4), guanine (5), adenine (6), thymine (7), cytosine (8), and uracil (9) molecules Calculations were made using. In these calculations, the interaction of molecules with proteins is seen as the most important factor that determines the activities of molecules (Lakhrissi et al., 2022; Tüzün, 2020). As the interactions of molecules with proteins increase, the activities of the molecules increase, which are generally chemical interactions, which are hydrogen bonds, polar and hydrophobic interactions, $\pi$-$\pi$ and halogen bonds (Celebioglu et al., 2021; Karrouchi et al., 2021; Majumdar et al., 2021; Riaz et al., 2021).

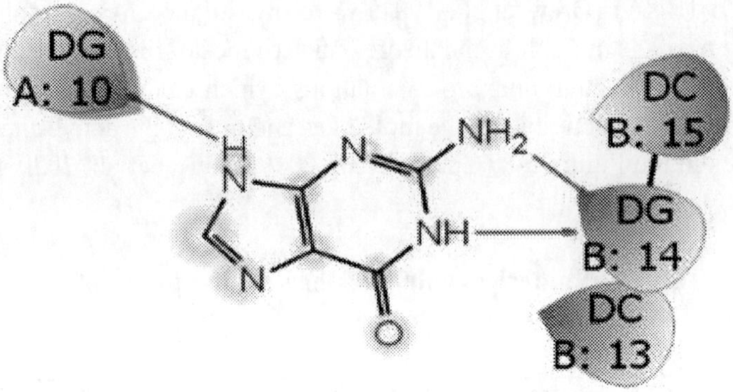
DG
A: 10
NH2
NH
DC
B: 15
DG
B: 14
DC
B: 13

**Figure 5.** Presentation interactions of guanine with 1BNA protein.

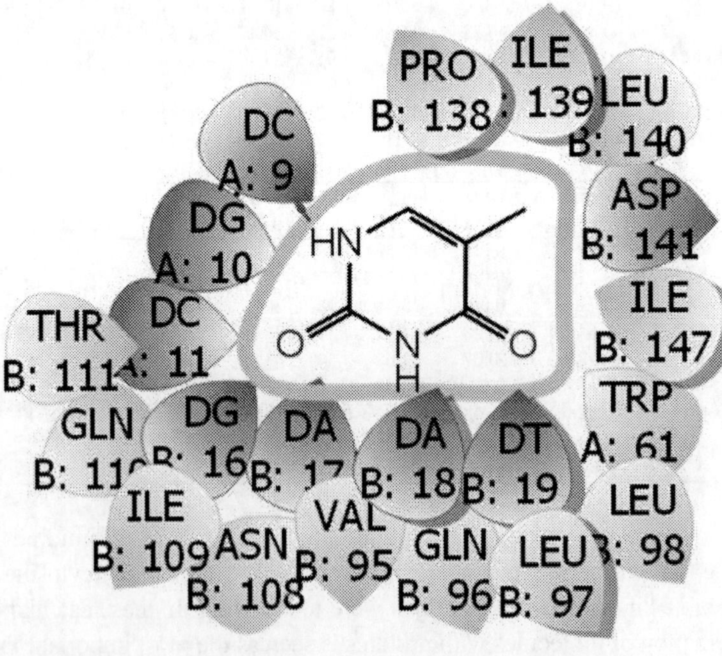
PRO
B: 138
ILE
139
LEU
B: 140
DC
A: 9
DG
A: 10
ASP
B: 141
ILE
B: 147
THR
B: 111
DC
11
HN
TRP
A: 61
GLN
DG
16
DA
DA
B: 18
DT
B: 19
LEU
98
ILE
B: 109
ASN
B: 108
VAL
B: 95
GLN
B: 96
LEU
B: 97

**Figure 6.** Presentation interactions of thymine with 1BNA protein.

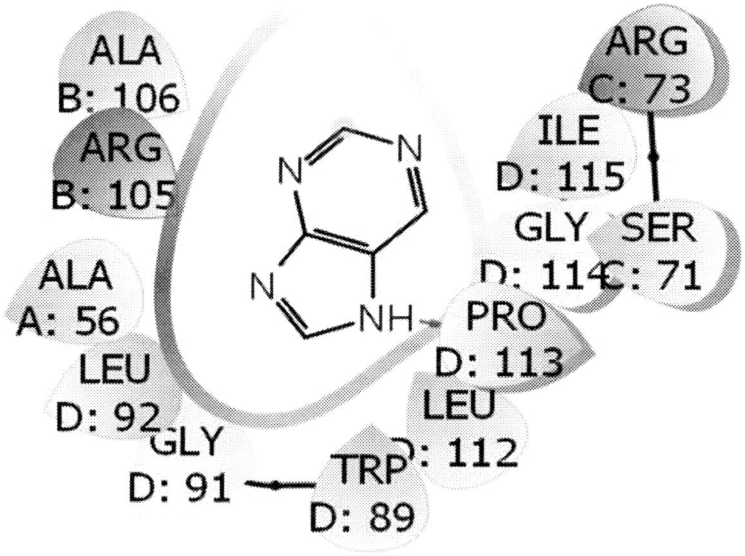
ALA
B: 106
ARG
B: 105
ALA
A: 56
LEU
D: 92
GLY
D: 91
TRP
D: 89
LEU
PRO
D: 113
ILE
D: 115
GLY
SER
C: 71
ARG
C: 73
N
N
N
NH

**Figure 7.** Presentation interactions of purine with 5YOD protein.

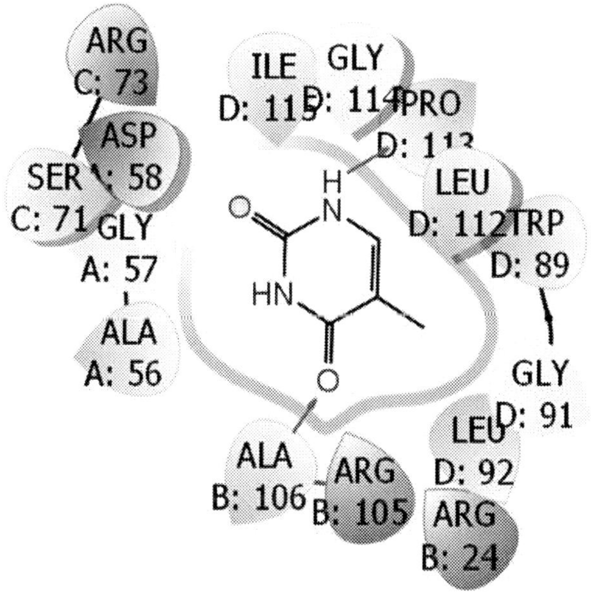
ARG
C: 73
ASP
SER
C: 71
GLY
A: 57
ALA
A: 56
ILE
GLY
PRO
LEU
D: 112
TRP
D: 89
GLY
D: 91
LEU
D: 92
ALA
B: 106
ARG
B: 105
ARG
B: 24
O
H
N
HN
O

**Figure 8.** Presentation interactions of thymine with 5YOD protein

In the theoretical calculations made, it is seen that the most active molecule against the 1BNA ID protein, which is a B-DNA DODECAMER protein, is the thymine molecule. It is seen that thymine molecule is the most active molecule with a docking score of -4.99. On the other hand, it is seen that the thymine molecule is the most active molecule against the 5YOD protein. It is seen that thymine molecule is the most active molecule with a docking score of -6.11 in.

## Conclusion

Research on pyrimidines, which started with Brugnatelli's synthesis of pyrimidine derivatives in 1818 and continued with Pinner, who first used the word pyrimidine in 1885, maintains its agenda even though it continues in different fields today. Since pyrimidines are essential components of RNA and DNA, it has been a matter of curiosity how they formed. Many hypotheses have been proposed about which of these compounds, which play the role of the key to life, were formed first. In order to understand the formation of pyrimidines, which are in the structure of both RNA and DNA, we have mentioned on these hypotheses in our article. Although it is likely that primidines or pyrimidine precursors or derivatives were formed on early Earth, whether in hydrothermal vents or somewhere else with favorable conditions, the fact that these molecules have also been detected in meteorites and comets and this complicates things a little. However, every researcher presents serious evidence in their scientific studies to support the hypothesis they support. Since the hypothesis that is the most accepted and supported in the Early Earth is the "RNA World" hypothesis, we evaluated this further. Another hypothesis, based on the detection of pyrimidine and other organic compounds in meteorites and comets, had to be mentioned. We did not give much space to the other hypotheses from the pyrimidine point of view. As a result of theoretical calculations, the interaction of many Pyrimidine derivative molecules with DNA was investigated. As a result of these interactions, it was seen that the thymine molecule was the molecule with the highest interaction. It is seen that the molecule with the highest activity against both 1BNA protein and 5YOD protein is thymine.

## References


Alberts, B., Johnson, A., Lewis, J., Raff, M., Roberts, K., & Walter, P. (Eds.). (2002). Cell Chemistry and Biosynthesis, *Molecular Biology of the Cell* (4th ed.) (pp 136–151), Garland Science.

Arimura, Y., Ikura, M., Fujita, R., Noda, M., Kobayashi, W., Horikoshi, N., ... & Kurumizaka, H. (2018). Cancer-associated mutations of histones H2B, H3. 1 and H2A. Z. 1 affect the structure and stability of the nucleosome. *Nucleic acids research,* 46(19), 10007-10018.

Becker, S., Feldmann, J., Wiedemann, S., Okamura, H., Schneider, C., Iwan, K., Crisp, A., Rossa, M., Amatov, T., & Carell, T. (2019). Unified prebiotically plausible synthesis of pyrimidine and purine RNA ribonucleotides, *Science,* 366, 76–82.

Bernhardt, H. S. (2012). The RNA world hypothesis: the worst theory of the early evolution of life (except for all the others), *Biology Direct,* 7, 23.

Brown, D. J., Evans, R. F., Cowden, W. B., & Fenn, M. D. (Eds.). (1994). Introduction to Pyrimidine Chemistry, In E. C. Taylor, (Ed.), *The Pyrimidines,* Volume 52 (Chemistry of Heterocyclic Compounds: A Series Of Monographs) (pp 1-47). John Wiley & Sons, Inc.

Cafferty, B. J., & Hud, N. V. (2015). Was a Pyrimidine-Pyrimidine Base Pair the Ancestor of Watson-Crick Base Pairs? Insights from a Systematic Approach to the Origin of RNA, *Isr. J. Chem.,* 55, 891–905.

Celebioglu, H. U., Erden, Y., Hamurcu, F., Taslimi, P., Şentürk, O. S., Özmen, Ü. Ö., ... & Gulçin, İ. (2021). Cytotoxic effects, carbonic anhydrase isoenzymes, α-glycosidase and acetylcholinesterase inhibitory properties, and molecular docking studies of heteroatom-containing sulfonyl hydrazone derivatives. *Journal of Biomolecular Structure and Dynamics,* 39(15), 5539-5550.

Drew, H. R., Wing, R. M., Takano, T., Broka, C., Tanaka, S., Itakura, K., & Dickerson, R. E. (1981). Structure of a B-DNA dodecamer: conformation and dynamics. *Proceedings of the National Academy of Sciences, 78*(4), 2179-2183.

Fox, S. W., & Harada, K. (1961). Synthesis of uracil under conditions of a thermal model of prebiological chemistry, *Science,* 133(3468), 1923–4.

Gavette, J. V., Stoop, M., Hud, N. V., & Krishnamurthy, R. (2016). RNA–DNA Chimeras in the Context of an RNA World Transition to an RNA/DNAWorld, *Angew. Chem. Int. Ed.,* 55, 13204–13209.

Glavin, D. P., & Bada, J. L. (2004). Isolation of Purines and Pyrimidines From The Murchison Meteorite Using Sublimation, *Lunar and Planetary Science XXXV. Conference Paper.*

Günsel, A., Kırbaç, E., Tüzün, B., Erdoğmuş, A., Bilgiçli, A. T., & Yarasir, M. N. (2019). Selective chemosensor phthalocyanines for Pd2+ ions; synthesis, characterization, quantum chemical calculation, photochemical and photophysical properties. *Journal of Molecular Structure,* 1180, 127-138.

Günsel, A., Kobyaoğlu, A., Bilgicli, A. T., Tüzün, B., Tosun, B., Arabaci, G., & Yarasir, M. N. (2020). Novel biologically active metallophthalocyanines as promising antioxidant-antibacterial agents: Synthesis, characterization and computational properties. *Journal of Molecular Structure,* 1200, 127127.

Haldane, J. B. S. (1929). Origin of Life, *The Rationalist Annual,* 148, 3–10.

Hill, A., & Orgel, L. E. (2002). Synthesis of adenine from HCN tetramer and ammonium formate, *Origins of Life and Evolution of Biospheres,* 32, 99–102.

Joyce, G. F. (2002). The antiquity of RNA-based evolution, *Nature,* 418, 214–221.

Karrouchi, K., Fettach, S., Tüzün, B., Radi, S., Alharthi, A. I., Ghabbour, H. A., ... & Garcia, Y. (2021). Synthesis, crystal structure, DFT, α-glucosidase and α-amylase inhibition and molecular docking studies of (E)-N'-(4-chlorobenzylidene)-5-phenyl-1H-pyrazole-3-carbohydrazide. *Journal of Molecular Structure*, 1245, 131067.

Kumar, S., & Narasimhan, B. (2018). Therapeutic potential of heterocyclic pyrimidine scaffolds, *Chemistry Central Journal,* 12, 38.

Lakhrissi, Y., Rbaa, M., Tuzun, B., Hichar, A., Ounine, K., Almalki, F., ... & Lakhrissi, B. (2022). Synthesis, structural confirmation, antibacterial properties and bio-informatics computational analyses of new pyrrole based on 8-hydroxyquinoline. *Journal of Molecular Structure,* 1259, 132683.

Levy, M., & Miller, S. L. (1998). The stability of the RNA bases: Implications for the origin of life, *Proc. Natl. Acad. Sci., 95*(14), 7933–7938.

Madigan, M. T., Bender, K. S., Buckley, D. H., Sattley, W. M., & Stahl, D. A. (Eds.). (2021). Microbial Evolution and Genome Dynamics, *Brock Biology of Microorganisms* (pp 428–459), Pearson Education.

Majumdar, D., Tüzün, B., Pal, T. K., Saini, R. V., Bankura, K., & Mishra, D. (2021). Structurally diverse heterobimetallic Pb (II)-Salen complexes mechanistic notion of cytotoxic activity against neuroblastoma cancer cell: Synthesis, characterization, protein–ligand interaction profiler, and intuitions from DFT. *Polyhedron,* 210, 115504.

Martina, M. G., Giannessi, L., & Radi, M. (2023). Multicomponent Synthesis of Purines and Pyrimidines: From the Origin of Life to New Sustainable Approaches for Drug-Discovery Applications, *Eur. J. Org. Chem.,* 26, e202201288, 1 of 17.

Martins, Z. (2018). The Nitrogen Heterocycle Content of Meteorites and Their Significance for the Origin of Life, *Life,* 8, 28.

Nam, I., Nam, H. G., & Zare, R. N. (2018). Abiotic synthesis of purine and pyrimidine ribonucleosides in aqueous microdroplets, *PNAS,* 115(1), 36–40.

Oparin, A. I. (1924). *Proiskhozhdenie zhizni,* [*The Origin of Life*], Izd. Moskovskiy Rabochiy,

Moscow.

Oró, J. (1961). Comets and the Formation of Biochemical Compounds on the Primitive Earth, *Nature,* 190, pages 389–390.

Oró, J. (1963). Studies in experimental organic cosmochemistry, *Ann. N.Y. Acad. Sci.,* 108, 464–81.

Oró, J. (2004). *Comets and The Origin of Life on The Primitive Earth,* In J. Seckbach (Ed.), Origins (pp 551–567). Kluwer Academic Publishers.

Plankensteiner, K., Reiner, H., Rode, B. M. (2005). Catalytic effects of glycine on prebiotic divaline and diproline formation, *Peptides,* 26, 1109–1112.

Ponnamperuma C., & Gabel, N. W. (1968). Current Status of Chemical Studies On The Origin of Life, *Space Life Sciences,* 1, 64–96.

Riaz, M. T., Yaqub, M., Shafiq, Z., Ashraf, A., Khalid, M., Taslimi, P., ... & Gulcin, I. (2021). Synthesis, biological activity and docking calculations of bis-naphthoquinone derivatives from Lawsone. *Bioorganic Chemistry,* 114, 105069.

Ricardo, A., Carrigan, M. A., Olcott, A. N., & Benner, S. A. (2004). Borate minerals stabilize ribose, *Science,* 303(5655), 196.

Rimmer, P. B., & Shorttle, O. (2019). Origin of Life's Building Blocks in Carbon- and Nitrogen-Rich Surface Hydrothermal Vents, *Life,* 9, 12.

Saad, N. Y. (2018). A ribonucleopeptide world at the origin of life, *Journal of Systematics and Evolution,* 56(1), 1–13.

Sagan, C. (1961). On the Origin and Planetary Distribution of Life, *Radiation Res.,* 15(2), 174–192.

Schrödinger Release 2021-3a: *Maestro,* Schrödinger, LLC, New York, NY, 2021.

Schrödinger Release 2021-3b: *Protein Preparation Wizard*; Epik, Schrödinger, LLC, New York, NY, 2021; Impact, Schrödinger, LLC, New York, NY; Prime, Schrödinger, LLC, New York, NY, 2021

Schrödinger Release 2021-3c: LigPrep, Schrödinger, LLC, New York, NY, 2021.

Schrödinger Release 2021-3d: QikProp, Schrödinger, LLC, New York, NY, 2021.

Segré, D., Ben-Eli, D., Deamer, D. W., & Lancet, D. (2001). The Lipid World, *Origins of Life and Evolution of the Biosphere,* 31, 119–145.

Shahzadi, I., Zahoor, A. F., Tüzün, B., Mansha, A., Anjum, M. N., Rasul, A., ... & Mojzych, M. (2022). Repositioning of acefylline as anti-cancer drug: Synthesis, anticancer and computational studies of azomethines derived from acefylline tethered 4-amino-3-mercapto-1, 2, 4-triazole. *Plos one,* 17(12), e0278027.

Thaddeus, P. (2006). The prebiotic molecules observed in the interstellar gas, *Philos. Trans. R. Soc. Lond. B Biol. Sci,* 361(1474), 1681–1687.

Tirard, S. (2017). J. B. S. Haldane and The Origin of Life, Journal of Genetics, 96(5), 735–739.

Türkan, F., Taslimi, P., Abdalrazaq, S. M., Aras, A., Erden, Y., Celebioglu, H. U., ... & Gülçin, İ. (2021). Determination of anticancer properties and inhibitory effects of some metabolic enzymes including acetylcholinesterase, butyrylcholinesterase, alpha-glycosidase of some compounds with molecular docking study. *Journal of biomolecular structure and dynamics,* 39(10), 3693-3702

Tüzün, B. (2020). Investigation of pyrazoly derivatives schiff base ligands and their metal complexes used as anti-cancer drug. *Spectrochimica Acta Part A: Molecular and Biomolecular Spectroscopy,* 227, 117663.

Wächtershäuser, G. (1992). Groundworks for an evolutionary biochemistry: The iron-sulphur world, *Progress in Biophysics and Molecular Biology,* 58(2), 85–201.

Zubay, G., & Mui, T. (2001). Prebiotic synthesis of nucleotides, *Origins of Life and Evolution of Biospheres,* 31, 87–102.

## Chapter 4

# The Versatility of the Pyrimidine Ring: Synthesis, Reactions, and Advanced Applications

**Priti Jain***
**Kritika**
**and Himanshi**
Department of Pharmaceutical Chemistry, School of Pharmaceutical Sciences,
Delhi Pharmaceutical Sciences and Research University, (Government of NCT of Delhi),
Sector-3, Pushp Vihar, New Delhi, India

## Abstract

Pyrimidine, a heterocyclic compound, has attracted significant attention in the field of organic chemistry and medicinal sciences due to its versatile nature and wide range of applications. This chapter provides a comprehensive overview of pyrimidine chemistry, synthetic methods, reaction pathways, medicinal importance, and recent advancements in pyrimidine research. The chapter begins with an introduction to pyrimidine's structure. The chemistry section explores the various synthetic strategies employed for pyrimidine synthesis, such as condensation reactions, cyclization processes, and rearrangements. Notable methodologies, such as the Green synthetic approach, Microwave synthesis, and Pinner reaction, are discussed in detail. The chapter further delves into the diverse reactions of pyrimidine derivatives, encompassing alkylation, halogenation, oxidation, reduction, nitration, diazo coupling and Vilsmeier-Haack reactions. By outlining their wide spectrum therapeutic applications, such as their

* Corresponding Author's Email: pritijain@dpsru.edu.in, pmj,grv@gmail.com.

In: The Chemistry of Pyrimidine Derivatives
Editor: Barry Schneider
ISBN: 979-8-89113-563-5

anticancer, antiviral, antifungal, and antibacterial capabilities, pyrimidine compounds' medicinal value is made clear. Noteworthy examples of pyrimidine-based drugs are highlighted, emphasizing their mechanism of action and clinical relevance. In the final sections, recent research advances in pyrimidine chemistry and pharmacology are summarized. This includes the exploration of novel synthetic methodologies, the discovery of new biological targets, and the development of more potent and selective pyrimidine-based drugs. Overall, this comprehensive data provides a valuable resource for researchers, chemists, and medicinal scientists interested in the chemistry, synthesis, reactions, medicinal importance, and recent developments related to pyrimidine compounds.

**Keywords:** pyrimidine, chemistry, medicinal importance, novel synthetic methodologies, green synthetic approach, microwave synthesis, pyrimidine derivatives, biological importance, recent advances

## 1. Introduction

Heterocyclic Chemistry consists of at least half of all organic chemistry research worldwide. Heterocyclic structures form the basis of many agrochemical, medicinal, and veterinary products. Heterocyclic compounds are one of the most important classes of compounds, with half of the organic compounds bearing heterocyclic ring system (Tolba et al., 2022). The main objective of organic and medicinal chemistry is to design, synthesize and produce molecules that have value for the treatment of humans, or which have human therapeutic importance. Heterocyclic compounds with pyrimidine moiety are becoming significant due to their broad range of biological effects, which include antioxidant, anticancer, antiviral, anti-inflammatory, antihypertensive, antidepressant, and antiplatelet characteristics (Abdelghani et al., 2017).

There are several six-membered rings with two heteroatoms heterocyclic compounds with biological activity. A heterocyclic chemical molecule with two nitrogen atoms in the first and third positions of its six-member ring (as shown in Figure1), pyrimidine is similar to benzene and pyridine. It shares isomers with other diazine forms. The energy of the ring's pi electrons drops as the nitrogen atom count increases, making nucleophilic aromatic substitution easier while electrophilic aromatic substitution becomes more difficult.(Katritzky AR, Ramsden CA, Scriven EFV, 2008) Pyrimidine and pyridine have a lot in common. Pyrimidine is a naturally occurring component of nucleic acids (Figure 2) (Uracil, Thymine, Cytosine, Orotic Acid) (Panneer Selvam et al., 2012).

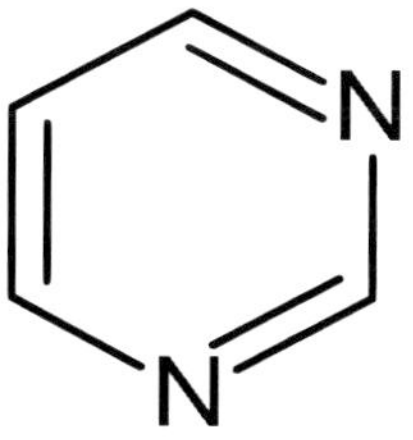

**Figure 1.** Pyrimidine ring structure.

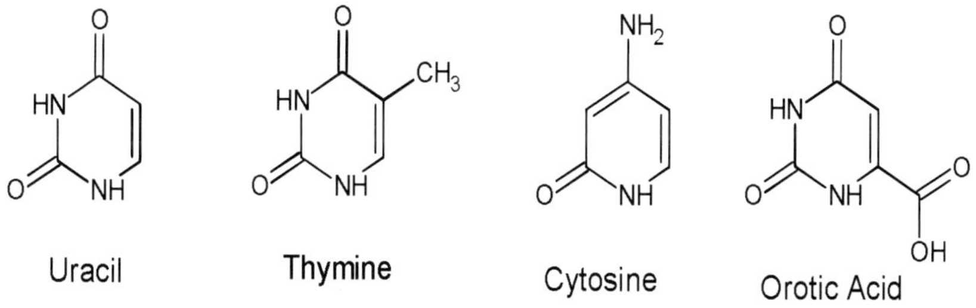
Uracil
Thymine
Cytosine
Orotic Acid

**Figure 2.** Structure of pyrimidine bases.

## 2. Source of Atoms in the Pyrimidine Ring

Pyrimidines were known as the products of uric acid. The first derivative isolated was alloxan (5,5- dihydroxy pyrimidine-2,4,6 (1H,3H,5H)-trione). Cytosine, uracil and thymine, the component of the chemical bases of DNA and RNA, make up pyrimidine (Naik & Chikhalia, 2007). A pyrimidine ring was formed using an amine and nitrogen (Figure 3) (M., L.I., 2005).

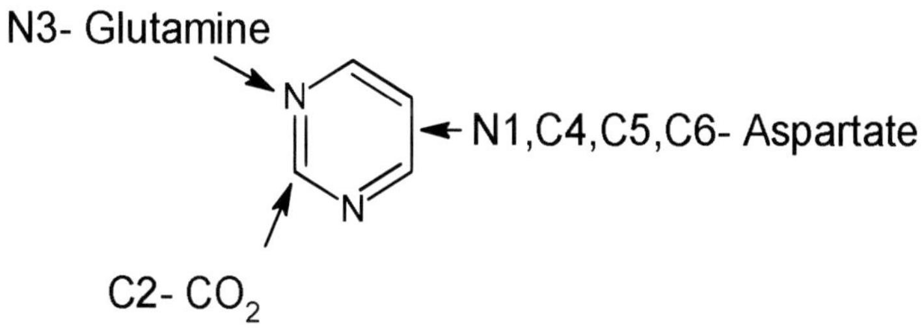
N3- Glutamine
N1,C4,C5,C6- Aspartate
C2- CO2

**Figure 3.** Source of atoms in pyrimidine ring.

- **N1, C4, C5 and C6** – Comes from Aspartate
- **N3** – Comes from Glutamine
- **C2**- Comes from Carbonic acid

## 3. Aromaticity of Pyrimidine

In 1931, Erich Huckel, a German chemist and physicist, introduced a fundamental principle known as Huckel's Rule. This rule is applied to determine whether an aromatic property of a planar ring molecule exists. According to Huckel's Rule, if a cyclic, planar molecule contains a total of $4n+2\pi$ electrons, it is considered aromatic. The proposition made by Erich Huckel has since become widely recognized and referred to as Huckel's Rule. (In,S., 1931)

**Huckel aromaticity rules are:**

Follow the checklist below to determine if the substance is aromatic. The compound is probably aromatic if it satisfies all of the following requirements.

- The molecule forms a closed cyclic structure.
- Each atom in the ring possesses a single p orbital.
- The molecule is planar, with sp2 hybridization across all atoms in the ring.
- It exhibits a closed loop of 4n+2 pi-bond electrons, where n represents any integer (0, 1, 2, 3, ...) (In, S., 1931).

Pyrimidine is a six-membered aromatic heterocycle. It consists of four carbon atoms and two nitrogen atoms, all of which are sp2 hybridized. Each of these atoms contributes a p orbital and a pi-electron, enabling pyrimidine to exhibit complete conjugation and aromaticity. While both nitrogen atoms in pyrimidine possess lone pair electrons in sp2 orbitals, these electrons are not involved in the aromatic system (Figure 4). This is confirmed by an electrostatic potential map of pyrimidine, which demonstrates that neither set of lone pair electrons is delocalized throughout the ring. (Terms, K.E.Y. 2019).

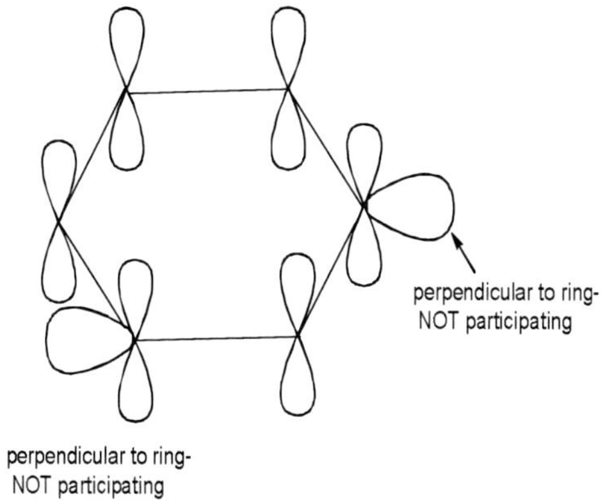
perpendicular to ring-
NOT participating
perpendicular to ring-
NOT participating

**Figure 4.** Orbital structure of pyrimidine.

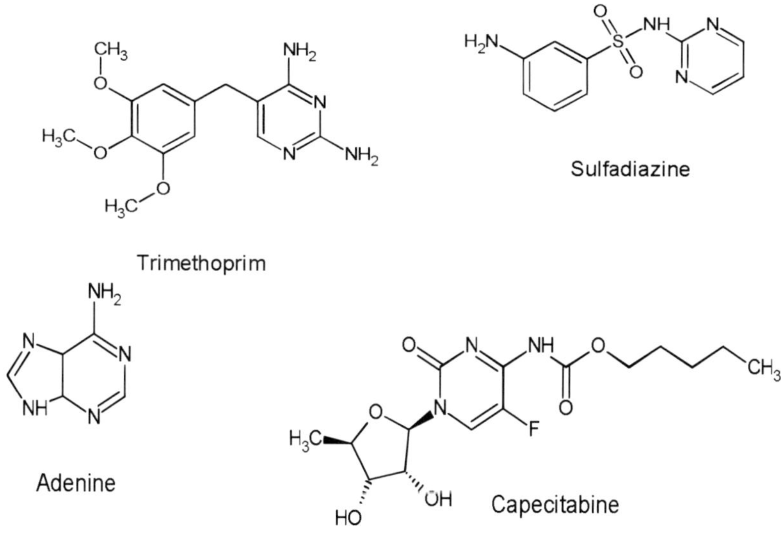
Trimethoprim
Sulfadiazine
Adenine
Capecitabine

**Figure 5.** Representative compounds containing a pyrimidine substructure.

# 4. Synthesis of Pyrimidine and Its Derivatives

In nature, the pyrimidine ring was produced by a four-step enzymatic mechanism that transforms glutamine, bicarbonate, and aspartate into orotate, a building block for the ribonucleotide biosynthesis process. Glutamine, ATP, and bicarbonate are converted into carbamoyl phosphate by carbamoyl phosphate synthetase II. The condensation of carbamoyl phosphate and aspartate was subsequently catalyzed by aspartate transcarbamoylase, yielding carbamoyl aspartate. Orotate, the precursor of ribonucleotides, was produced by oxidising orotate with dihydroorotate dehydrogenase after dihydroororotase promotes dehydration.(P. S. Baran, R. A. Shenvi, S. A. Nguyen, 2006) The pyrimidine ring, a heterocyclic molecule with two nitrogen atoms at positions 1 and 3 in the ring, was yielded during the synthesis of pyrimidine (Figure 5). There are various methods to synthesize pyrimidine, the choice of the synthesis method depends on the specific pyrimidine derivative you want to prepare and the starting materials available (Hill & Movassaghi, 2008).

### 4.1. Pinner-Type Synthesis

The German chemist Adolf Pinner, who originally published this technique in the late 19th century, is credited for discovering the Pinner reaction. The Pinner reaction, a historical chemical reaction, involves the condensation of amidines with 1,3-dicarbonyl compounds (Figure 6). Recent advancements in this area have focused on expanding the range of substitution patterns found in the resulting pyrimidines or on identifying new reaction conditions that enable the use of readily accessible starting materials (Mahfoudh et al., 2017).

### 4.2. Synthesis of Pyrimidines from (2,2,2 Trichloroethylene) Acetophenones and Amidines

Recently, trichloromethyl-substituted α and β-unsaturated ketones were utilized as effective substrates for condensation reactions with amines. The trichloromethyl group serves as an effective traceless activating group that was eliminated during the final re-aromatization step. While using a stepwise process enables the isolation of each reaction intermediate - 3 and 4 leads directly to 2,4-disubstituted pyrimidines in good yields. The starting (2,2,2trichloroethylene)acetophenones were obtained by reacting the corresponding acetophenones with chloral (Figure 7) (Guirado et al., 2016).

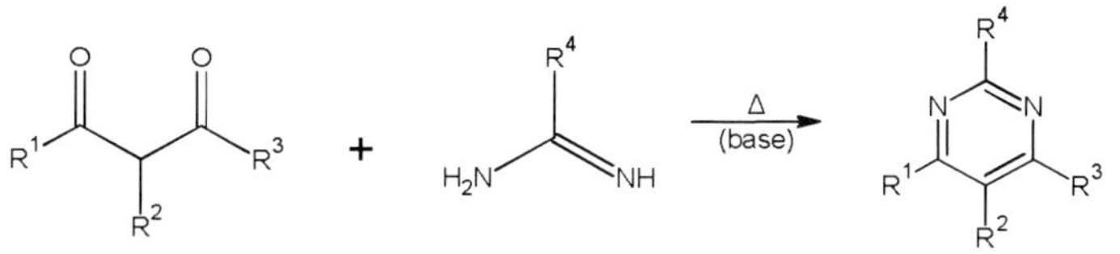
O
O
R1
R3
R2
+
R4
H2N
NH
Δ
(base)
R4
N
N
R1
R3
R2

**Figure 6.** Pinner synthesis.

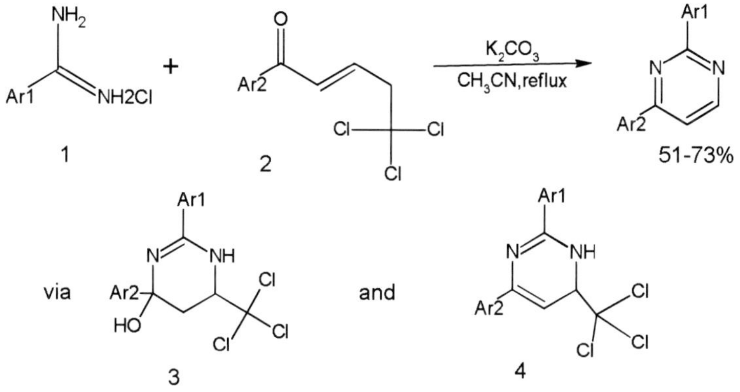
NH2
Ar1
NH2Cl
1
+
O
Ar2
Cl
Cl
Cl
2
K2CO3
CH3CN,reflux
Ar1
N
N
Ar2
51-73%
via
Ar1
N
NH
Ar2
HO
Cl
Cl
Cl
3
and
Ar1
N
NH
Ar2
Cl
Cl
Cl
4

**Figure 7.** Pyrimidines from acetophenones and amidines.

## 4.3. Sequential Suzuki Coupling/Condensation Reaction

A similar approach is based on sequential condensation/cross-coupling functionalization. In this case, 3-iodochromone was initially involved in a Suzuki coupling reaction with various aryl boronic acids, followed by condensation with acetamidine or guanidine to produce the corresponding 2-amino- or 2-methyl-4-(2-hydroxyphenyl)5-aryl pyrimidines, yielding 35-85% yields (Figure 8) (Jiao et al., 2015).

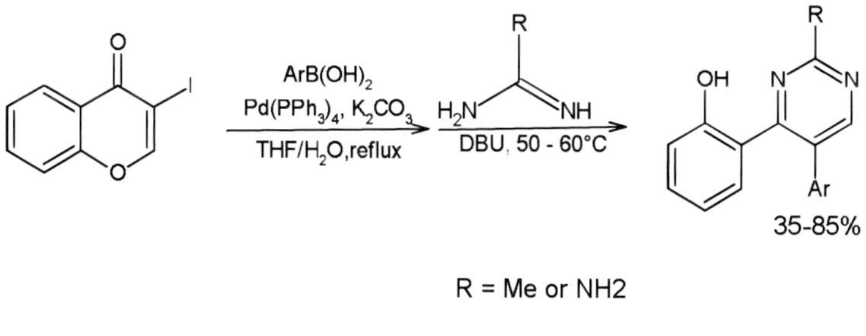
O
I
O
ArB(OH)2
Pd(PPh3)4, K2CO3
THF/H2O,reflux
R
H2N
NH
DBU, 50 - 60°C
R
OH
N
N
Ar
35-85%
R = Me or NH2

**Figure 8.** Sequential Suzuki coupling/condensation reaction.

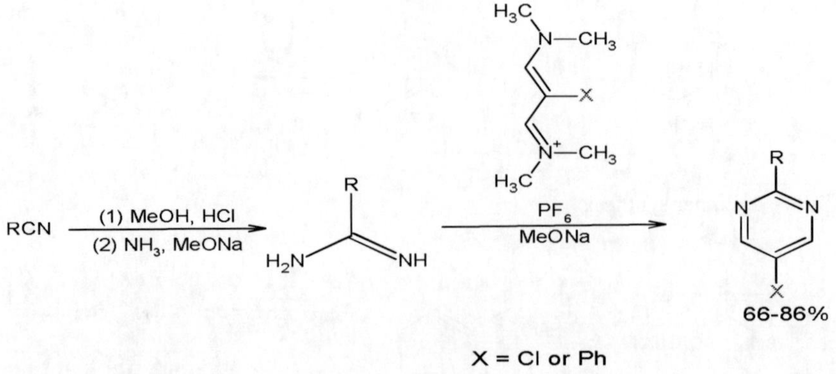
H3C
N—CH3
X
N+—CH3
H3C
R
RCN
(1) MeOH, HCl
(2) NH3, MeONa
H2N
NH
PF6
MeONa
R
N
N
X
66-86%
X = Cl or Ph

**Figure 9.** Green one-pot synthesis of pyrimidine from nitriles.

## 4.4. Green One-Pot Synthesis of Pyrimidine from Nitriles

Frutos and Wei described an original one-pot synthesis from nitriles. The nitrile undergoes conversion into the corresponding amidine via a classical strategy, and this was then directly condensed with 3-chloro- or 3-phenylvinamidinium salts to produce the corresponding 2,5-disubstituted pyrimidines in good yields (Figure 9). However, the range of the reaction was constrained by the usage of vinamidinium salts as 1,3-dicarbonyl substitutes (Frutos et al., 2013).

## 4.5. Green Synthesis of Phthalide Fused Pyrimidine

The field of organic synthesis has witnessed significant interest in the development of new, environmentally friendly methods. One particular area of focus is the utilization of water as a solvent, which offers numerous advantages. Water is a desirable option from an economic and environmental standpoint because it is easily accessible, affordable, non-toxic, and inflammable. A reaction's outcome can be significantly influenced by the special structure and physicochemical characteristics of water, such as its polarity, hydrogen bonds, hydrophobic effects, and trans-phase interactions (Lindstrom, 2007).

Phthalide is an important building block that may be found in many natural products and has exceptional biological properties. Phthalides have advantageous biological effects when functionalized at C-3 (Arcelo, M.;

Ravina, E.; Masaguer, C. F.; Dominguez, E.; Areias, F. M.; Brea & Loza, 2007). Additionally, bioactive phthalides such as the phytotoxic and cytotoxic compounds serve as further examples of their biological significance (Shi, X.-Y.; Li, 2012).

Considerable interest has been shown in the idea of combining two pharmacophores on a common scaffold to create hybrid molecules or conjugates. In order to produce a chemical entity that demonstrates greater medical efficacy than its constituents, these hybrids combine two active components into a single molecule. Due to this method, phthalide-fused pyrazoles or pyrimidines, which combine two physiologically active heterocyclic cores, have been investigated (Figure 10). These compounds show promise for pharmacological research (Motamedi et al., 2014), (Antonello, A.; Hrelia, P.; Leonardi, A.; Marucci, G.; Rosini, M.; Tarozzi & Tumiatti, V.; Melchiorre, 2005).

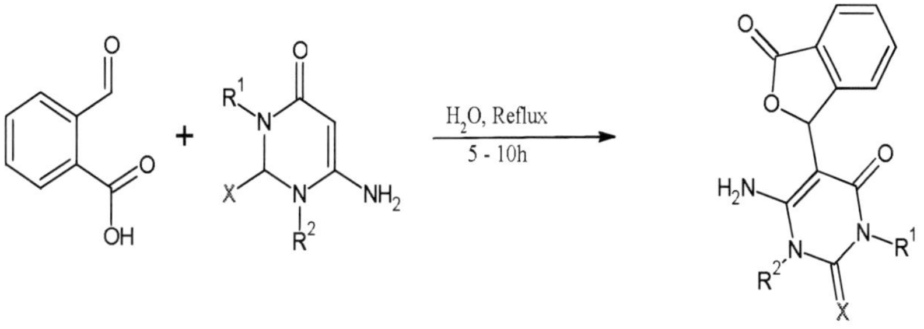
O
OH
O
+
R1
N
O
X
N
R2
NH2
H2O, Reflux
5 - 10h
O
O
H2N
O
N
R1
R2
N
X

**Figure 10.** Green synthesis of Phthalide fused pyrimidine.

**Table 1.** Comparison between different derivatives synthesized by the green method (Motamedi et al., 2014)

| **R¹** | **R²** | **X** | **Time(h)** | **Yield (%)** |
|---|---|---|---|---|
| Me | Me | O | 6 | 92 |
| H | H | O | 10 | 80 |
| H | Me | O | 7 | 86 |
| H | H | S | 9 | 74 |
| H | _ | SMe | 5 | 89 |

(Motamedi et al., 2014)

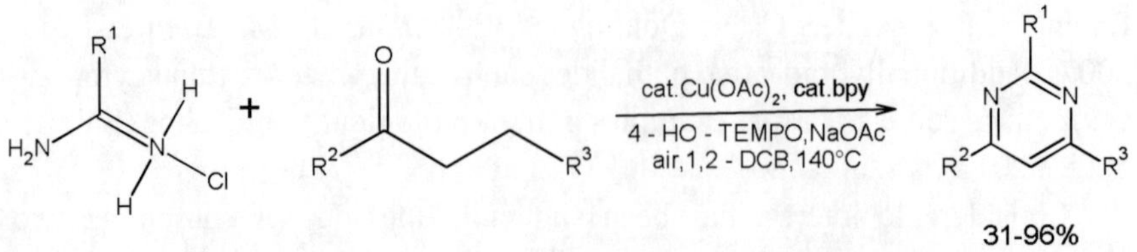
cat.Cu(OAc)2, cat.bpy
4 - HO - TEMPO,NaOAc
air,1,2 - DCB,140°C
31-96%

**Figure 11.** Oxidative copper-catalyzed condensation of ketones and amidines.

## 4.6. Oxidative Copper-Catalyzed Condensation of Ketones and Amidines

Han recently reported a novel approach in which enones are generated *in situ* from ketones using oxidative copper-catalyzed conditions and then condensed with amidines to produce the corresponding 2,4,6-trisubstituted pyrimidines (Figure 11). The oxidation of the starting ketone to the corresponding enone is likely to occur via a single-electron transfer from a copper enolate. This reaction is quite general, with a broad range of propiophenones and aliphatic ketones being well-tolerated, and it is equally effective with both aromatic and aliphatic amidines (Dang et al., 2018).

## 4.7. Kemp's Three-Component Synthesis of Pyrimidine

In 2015, Kempe reported a novel three-component condensation method that involves primary alcohols, secondary alcohols, and amidines under iridium catalysis conditions to yield 2,4,6-trisubstituted and 2,4,5,6-tetrasubstituted pyrimidines. The reaction proceeds through two successive dehydrogenations of the two alcohols, followed by a base-mediated aldol reaction and by dehydrative amidine condensation onto the resulting enone (Figure 12). This strategy demonstrates a wide scope of reactivity and remarkable tolerance to both aliphatic and aromatic substituents, resulting in good yields (Deibl et al., 2015).

## 4.8. Kemp's Four-Component Synthesis of Pyrimidine

An advancement in molecular diversity was made through a consecutive four-component process utilizing the same iridium catalyst. Initially, secondary alcohol and primary alcohol react, resulting in the formation of an elongated secondary alcohol, which serves as the source for the C-4 and C-5 substituents. Subsequently, primary alcohol and the amidine compound are introduced to complete the synthesis of pyrimidines with substitutions at positions 2, 4, 5 and 6 (Figure 13). The reaction exhibits good yields and compatibility with a wide range of substitution patterns (Zahedifar & Sheibani, 2015).

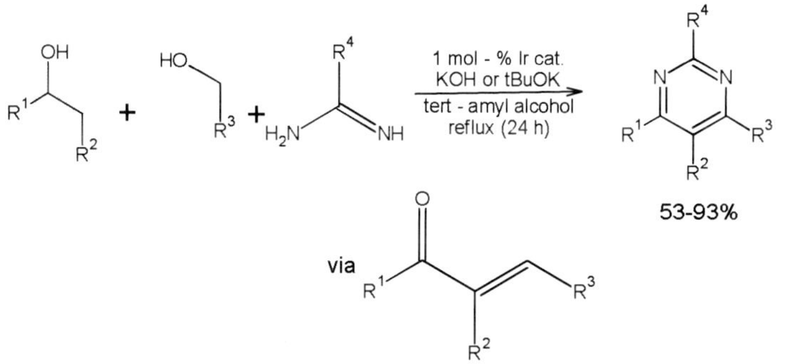
OH
HO
R4
1 mol - % Ir cat.
KOH or tBuOK
tert - amyl alcohol
reflux (24 h)
R1
R2
R3
H2N
NH
N
53-93%
O
via

**Figure 12.** Kemp's three-component synthesis of pyrimidine.

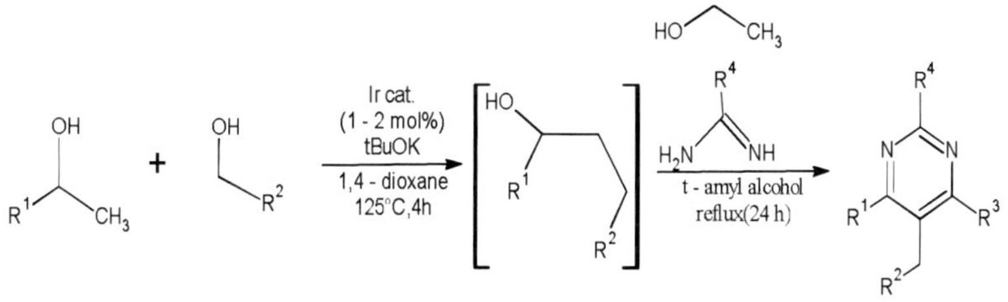
HO
CH3
OH
Ir cat.
(1 - 2 mol%)
tBuOK
1,4 - dioxane
125°C,4h
R4
H2N
NH
t - amyl alcohol
reflux(24 h)
R1
R2
R3
N

**Figure 13.** Kemp's four-component synthesis of pyrimidine.

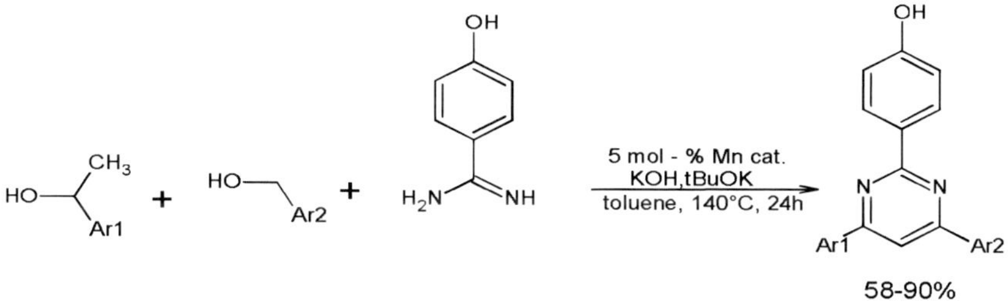
OH
CH3
HO
Ar1
Ar2
H2N
NH
5 mol - % Mn cat.
KOH,tBuOK
toluene, 140°C, 24h
N
58-90%

**Figure 14.** Kirchner three-component synthesis of pyrimidines.

## 4.9. Kirchner Three Component Synthesis of Pyrimidines

Kirchner has reported on a method similar to Kempe's, using a manganese-based catalyst, for three-component condensation reactions with primary alcohols, secondary alcohols, and amidines. The reaction starts with methyl carbinol, primary alcohol, and benzamidine under almost identical conditions as Kempe's method, resulting in the formation of the corresponding pyrimidines in good yields (Figure 14) (Mastalir et al., 2016).

## 4.10. Diels – Alder Cycloaddition between 1,2,3-Triazine and Amidines

The Boger group conducted a study on Diels-Alder cycloaddition between a range of 1,2,3-triazines and electron-rich dienophiles in 2011. Among the dienophiles, aliphatic and aromatic amidines showed fast reactions resulting in pyrimidines with yields greater than 90%. The reactions were initiated with unsubstituted 1,2,3-triazine and alkyl (R = Me) aromatic, or heteroaromatic amidines under mild heating conditions. The final ammonia elimination step was rate-limiting and required heating (Figure 15). The [4+2] cycloaddition and subsequent cyclo-reversion occurred within minutes, even at low temperatures in some cases. The use of imitates instead of amidines was also effective but resulted in lower yields. The amidines should be used as their free bases rather than their HCl salts to ensure clean reactions and high yields (Boger, 1986).

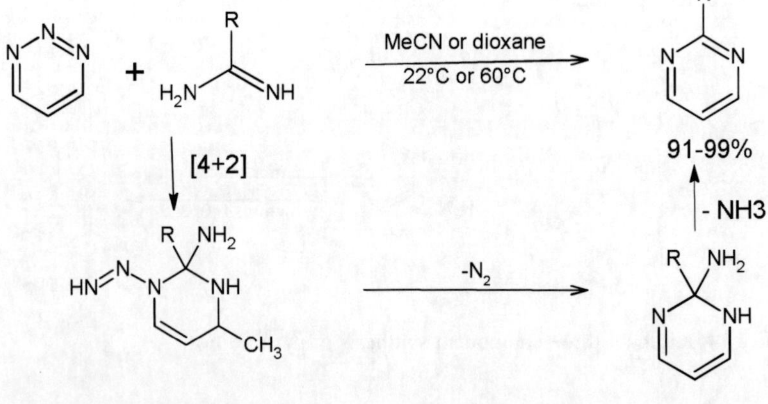
R
MeCN or dioxane
22°C or 60°C
+
$H_2N$
NH
N
R
91-99%
[4+2]
- NH3
R NH$_2$
HN
NH
CH$_3$
-N$_2$
R NH$_2$
NH

**Figure 15.** Diels – Alder cycloaddition between 1,2,3-triazine and amidines.

## 4.11. Microwave Method for Synthesis

The microwave method has emerged as the most valuable method for synthesis alternative to conventional heating methods in recent years. Microwave-assisted chemistry offers several advantages over traditional methods, such as enhanced efficiency, faster heating rates, and the ability to quickly optimize synthetic procedures. By utilizing microwaves as an energy source, direct contact between the chemical reactants and the energy source is not required. This results in improved energy usage and often leads to increased reaction rates and overall efficiency. In recent times, there have been significant advancements in instrumentation for microwave chemistry. The introduction of dedicated ovens designed specifically for organic synthesis, which focuses on microwaves in a monomodal cavity, has greatly contributed to the popularity and reproducibility of microwave-assisted chemistry. These advancements have expanded the range of methodologies available for the development of new synthetic reactions and the optimization of existing procedures (Bagley et al., 2003), (Bagley, M. C.; Dale, 2002).

### *4.11.1. Preparation Methods*

Chalcone derivatives have been synthesized through the condensation of different substituted aryl aldehydes and acetophenone in alkaline ethanol (Figure 16). Additionally, pyrimidine-2-one derivatives have been obtained by combining chalcones with urea using both conventional and ultrasonic methods. The chalcones were synthesized through the Claisen-Schmidt condensation technique, employing both conventional and ultrasonic-assisted methods (Shilpa et al., 2012).

The synthesis of 2-amino-3-ethylcarboxylate-4,5-diphenylpyrroles was conducted using the following procedures: 2-amino-2-phenylacetophenones and cyano-ethyl acetate were dissolved in 10 ml of ethanol (EtOH), and the resulting solution was adsorbed over 20g of basic alumina or montmorillonite. The reaction mixture was then placed in a beaker and subjected to microwave irradiation in an alumina bath for 6-7 minutes intermittently.

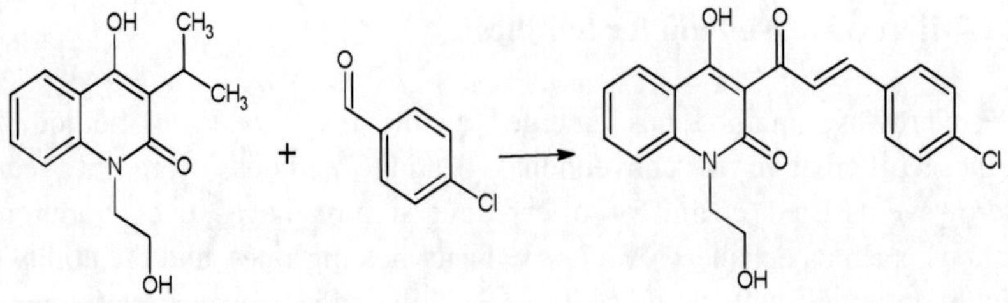
OH
CH3
CH3
N
O
OH
O
Cl
+
OH
O
N
O
Cl
OH

**Figure 16.** Chalcone synthesis.

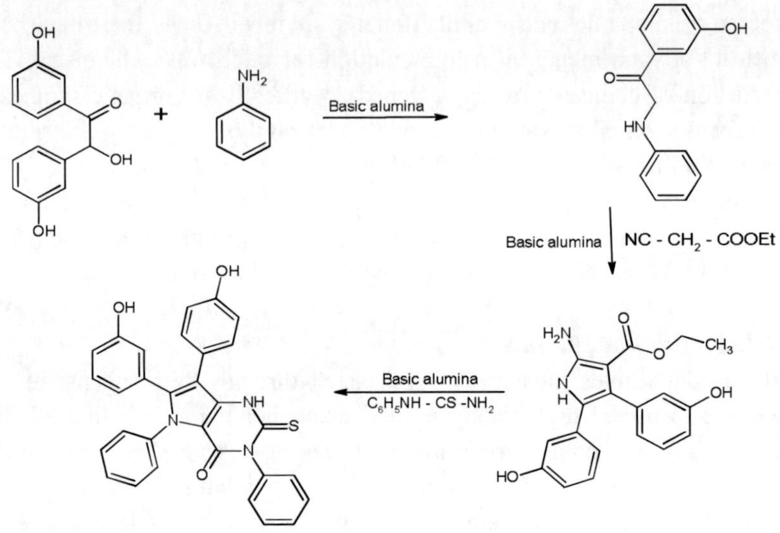
OH
O
OH
OH
+
NH2
Basic alumina
OH
O
HN
Basic alumina
NC - CH2 - COOEt
H2N
O
O
CH3
N
H
OH
HO
Basic alumina
C6H5NH - CS -NH2
OH
OH
NH
N
S
O
N

**Figure 17.** Synthesis of 2-thioxo-3,7-disubstituted-5,6-diphenylpyrrolo[2,3-d] pyrimidin-4(1H)-ones.

The resultant substance is then subjected to further treatment with basic alumina to produce 2-thioxo-3,7-disubstituted-5,6-diphenylpyrrolo[2,3-d] pyrimidin-4(1H)-ones (Figure 17).

Both conventional and microwave synthesis methods were used (Kidwai & Mishra, 2004).

The synthesis of 3,4-dihydro benzo 2,3-d pyrimidines by cyclization of 1,3-cyclohexadiene derivatives with formamide under acidic conditions in dry media under microwave irradiations (Figure 18) (Safaei-Ghomi & Ghasemzadeh, 2011).

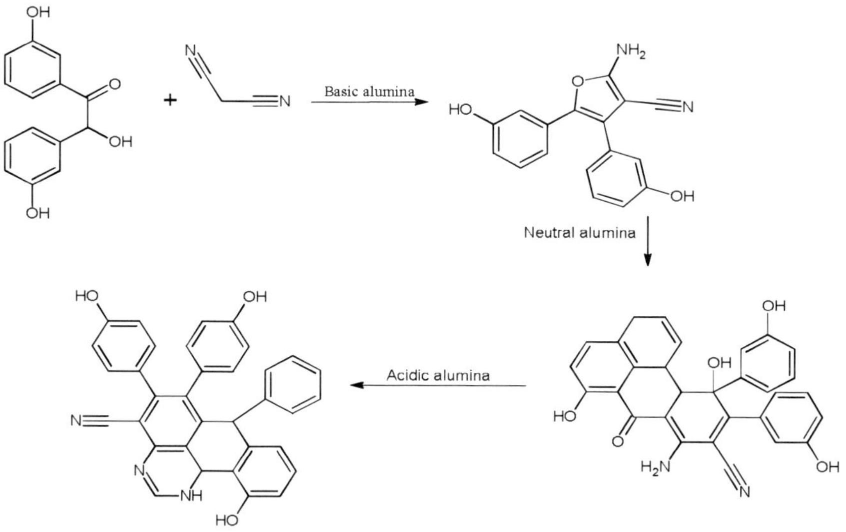
Basic alumina
Neutral alumina
Acidic alumina

**Figure 18.** Synthesis of 3,4-dihydro benzo 2,3-d pyrimidines.

**Table 2.** Comparison between microwave-assisted and conventional methods of synthesis (Shilpa et al., 2012)

| Structure | Microwave | irradiation | Conventional Heating | |
|---|---|---|---|---|
| | Time (min) | Yield (%) | Time (hour) | Yield (%) |
| 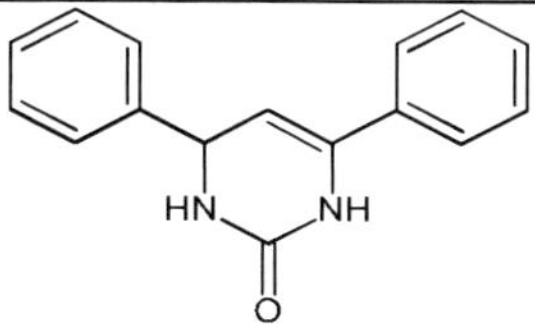 | 24 | 82 | 6.5 | 65 |
| 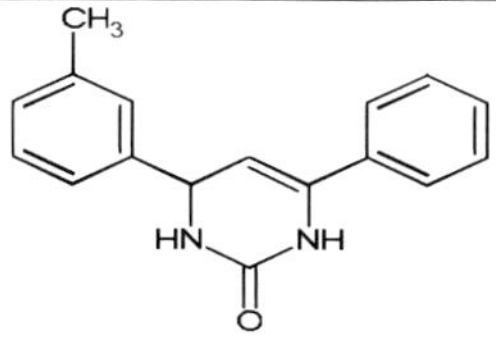 | 21 | 80 | 6 | 54 |
| 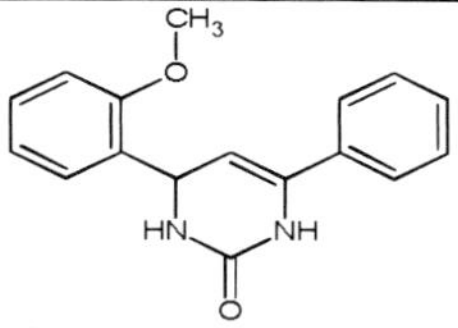 | 25 | 75 | 5.5 | 60 |

**Table 2.** (Continued)

| Structure | Microwave irradiation | | Conventional Heating | |
|---|---|---|---|---|
| | **Time (min)** | **Yield (%)** | **Time (hour)** | **Yield (%)** |
| 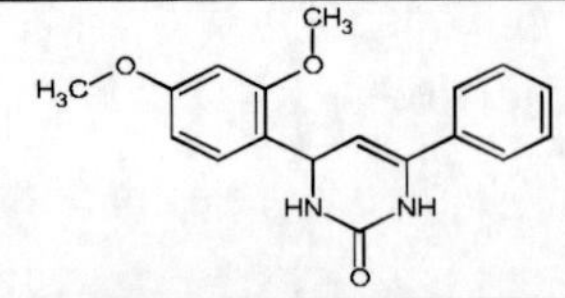 | 22 | 73 | 5 | 65 |
| 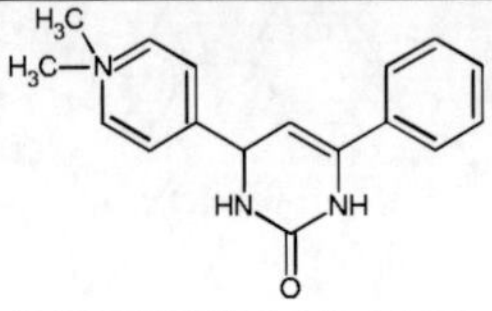 | 29 | 75 | 5.5 | 55 |
| 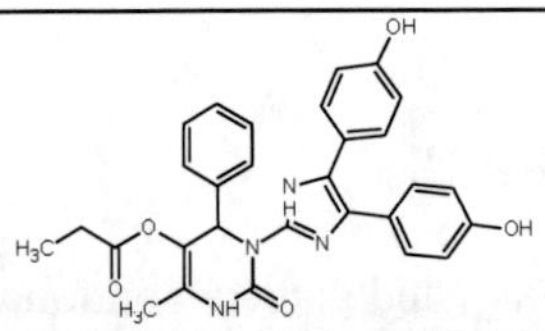 | 8 | 79 | 12 | 64 |
| 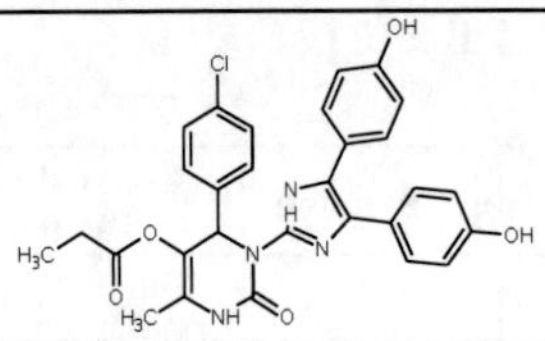 | 8 | 69 | 12 | 62 |
| 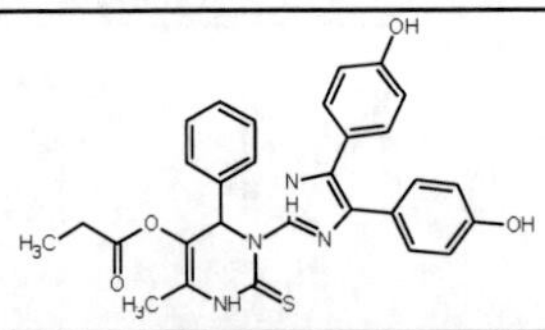 | 8 | 78 | 12 | 65 |
| 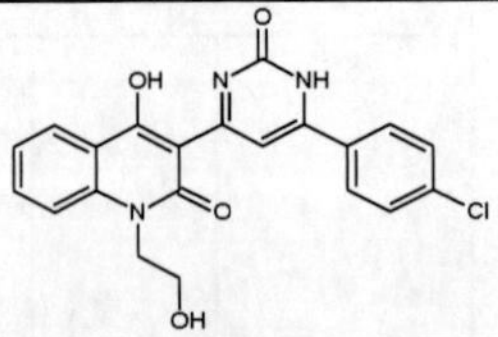 | 4.5 | 85 | 10 | 40 |
| 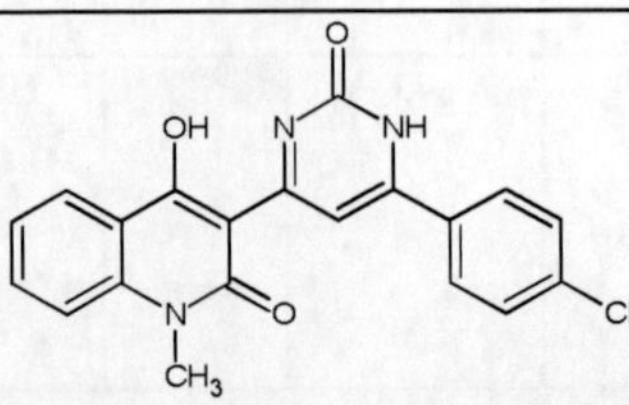 | 4.5 | 82 | 10 | 39 |

| Structure | Microwave | irradiation | Conventional Heating | |
|---|---|---|---|---|
| | Time (min) | Yield (%) | Time (hour) | Yield (%) |
| 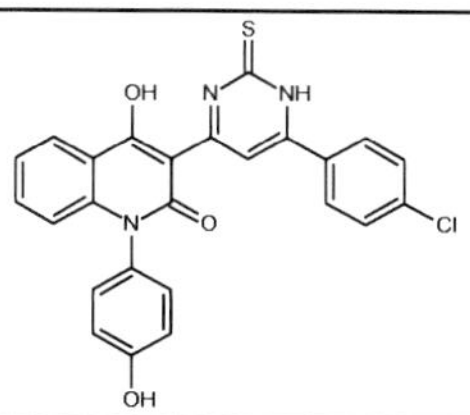 | 4.0 | 82 | 8 | 38 |
| 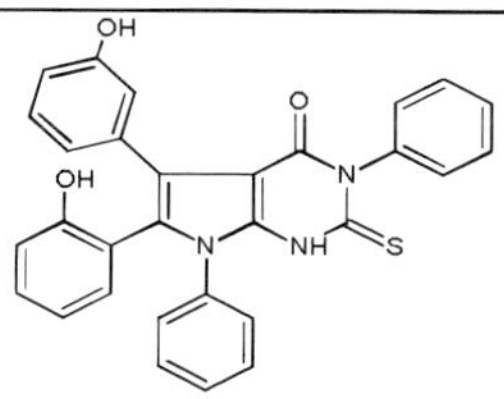 | 9 | 80 | 12 | 62 |
| 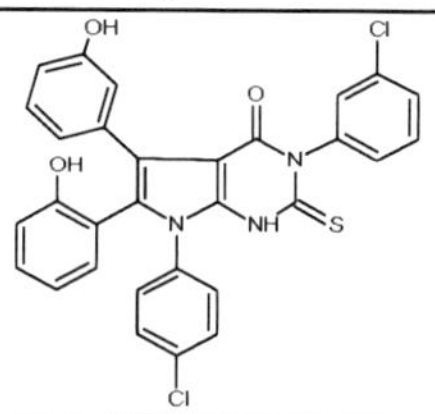 | 8 | 90 | 11.5 | 65 |
| 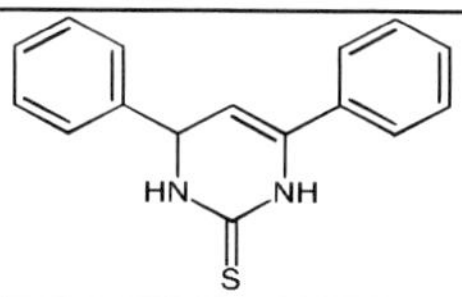 | 20 | 82 | 5.5 | 65 |
| 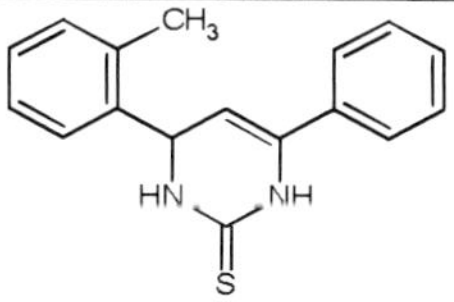 | 22 | 78 | 6 | 55 |
| 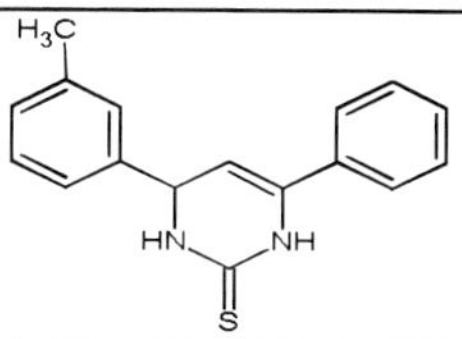 | 22 | 80 | 6 | 54 |
| 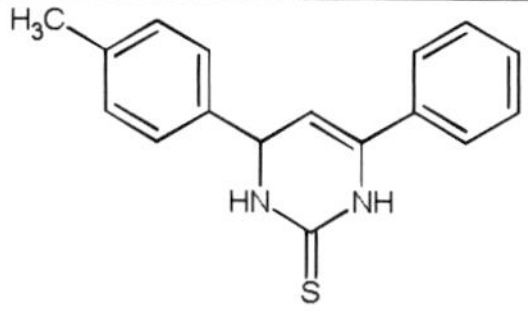 | 24 | 76 | 5.5 | 58 |

(Shilpa et al., 2012)

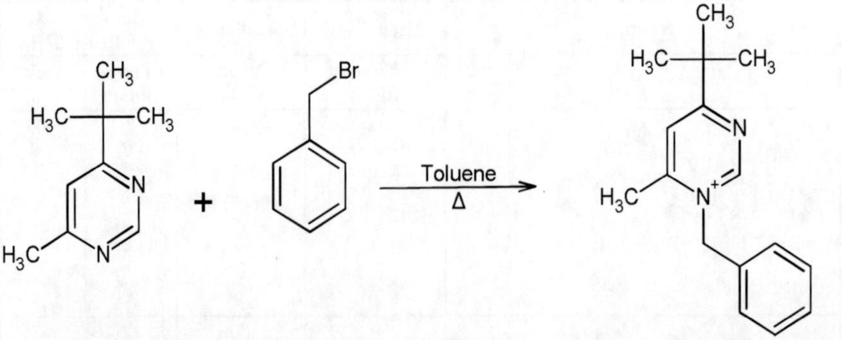
Br
Toluene
Δ

4-t-butyl-6-methyl pyrimidine

1-benzylated product

**Figure 19.** Alkylation reaction of pyrimidine derivatives at the nitrogen.

## 5. Reactions of Pyrimidine

### 5.1. Alkylation Reaction

Pyrimidine undergoes an alkylation reaction in which an alkyl group is added to the pyrimidine ring. The alkylation of pyrimidine can occur at various positions on the ring, resulting in different alkylated derivatives. The alkylation reaction can be carried out using different conditions and methods depending on the specific alkylating agent and desired product. Studies have shown that electrophiles can react with nitrogen atoms present in aromatic rings. In the case of pyrimidines with non-tautomerizable substituents, the process of simple alkylation is largely influenced by steric factors. An example of this is observed in the reaction of 4-t-butyl-6-methyl pyrimidine with benzyl chloride in toluene, where only 1-benzylated product is formed (Figure 19) (Curphey & Prasad, 1972).

### 5.2. Oxidation

Low yields of N-oxides are produced by direct N-oxidation on pyrimidines and methyl pyrimidine due to their susceptibility to breakdown, ring-carbon oxidation, and ring-opening reactions. There must be Activating Substituents. Figure 20 shows that when pyrimidine was combined with m-chloroperbenzoic acid in chloroform, pyrimidine N-oxides were produced in a yield of 48 percent, however when 2-methyl pyrimidine was employed as the starting material, 2-methyl pyrimidine N-oxides were produced in a yield of 55 percent (Undheim, 1996).

## 5.3. Reduction

$NaB(CN)H_3$ can be used to reduce pyrimidine in methanol while trapping the reduced forms with benzyl chloroformate to create pyrimidine enamine, which is dibenzyl pyrimidine-1,3 (2H,4H) -dicarboxylate (Figure 21) (Undheim, 1996).

## 5.4. Nitration

Pyrimidine and its cation possess a high degree of $\pi$-deficiency, making them resistant to nitration. The reaction is made more difficult by the electron-depleted state of the $\pi$ -system in the 5-nitro derivative. Adducts are thought to be created during nitration and may go through oxidation or ring-opening processes. Nitration of pyrimidine is a challenging task. However, Pyrimidines that are aryl groups substituted are frequently nitrated preferentially at the aryl site. As shown in the Figure 22, nitration of 4-phenyl pyrimidine in the presence of a concentrated mixture of nitric and sulfuric acids resulted in the production of 40% 4-o-nitrophenyl pyrimidine and 60% 4-m-nitrophenyl pyrimidine (Kulkarni, 2014).

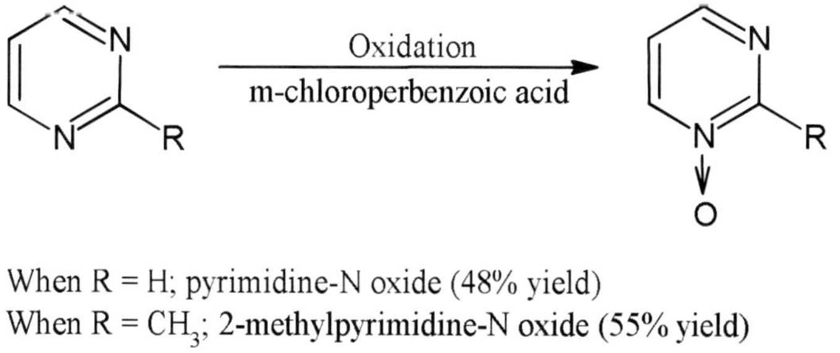
N
N
R
Oxidation
m-chloroperbenzoic acid
N
N
R
O
When R = H; pyrimidine-N oxide (48% yield)
When R = CH3; 2-methylpyrimidine-N oxide (55% yield)

**Figure 20.** Oxidation reaction of 2-substituted pyrimidine at the nitrogen.

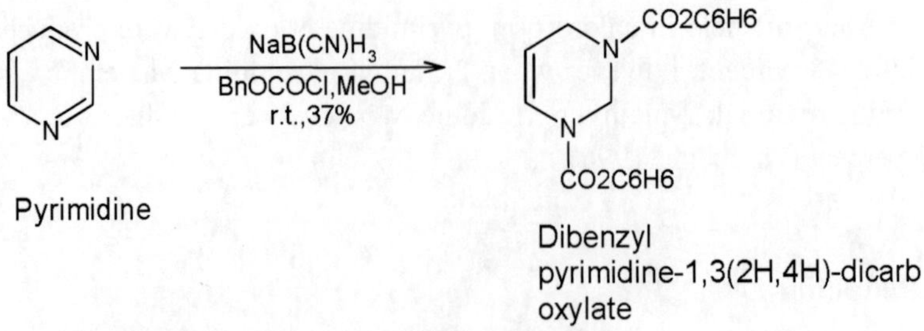
NaB(CN)H3
BnOCOCl,MeOH
r.t.,37%
CO2C6H6
CO2C6H6
Pyrimidine
Dibenzyl
pyrimidine-1,3(2H,4H)-dicarb
oxylate

**Figure 21.** Reduction of pyrimidine by $NaB(CN)H_3$.

## 5.5. Halogenation Reaction

Pyrimidines can be directly halogenated in the 5-position with the use of electrophilic reagents. Nucleophilic exchange reactions occur in electrophilic positions. Electron-donating groups, such as hydroxyl, amino, or 2-tert-butyl groups, must be used to activate pyrimidine in the 5-position before chlorination can occur. Suitable chlorinating agents include chlorine in the presence of a base, phenyl iododichloride, thionyl chloride, or sulfuryl chloride with ferric chloride as a catalyst. As depicted in the Figure 23, 4-amino-2-hydroxypyrimidine was treated with sulfuryl chloride in the presence of ferric chloride to get 4-amino-5-chloro-2-hydroxypyrimidine. Bromine can produce 71-78 percent of 4-amino-5-bromo-2-hydroxypyrimidine in solvents like benzene or nitrobenzene (Figure 24) (Jubeen et al., 2018).

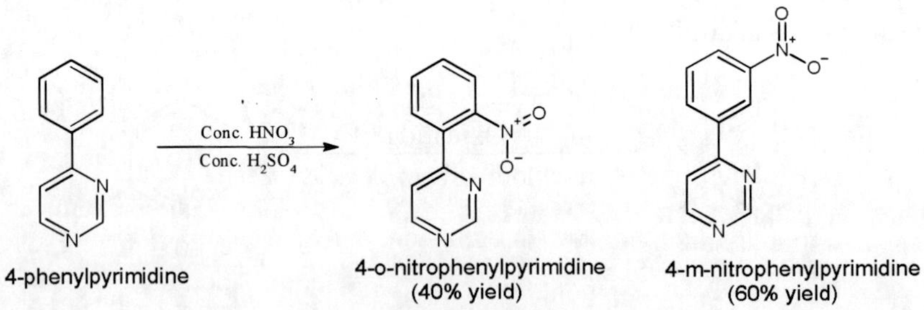
Conc. HNO3
Conc. H2SO4
4-phenylpyrimidine
4-o-nitrophenylpyrimidine
(40% yield)
4-m-nitrophenylpyrimidine
(60% yield)

**Figure 22.** Nitration of 4-phenylpyrimidine.

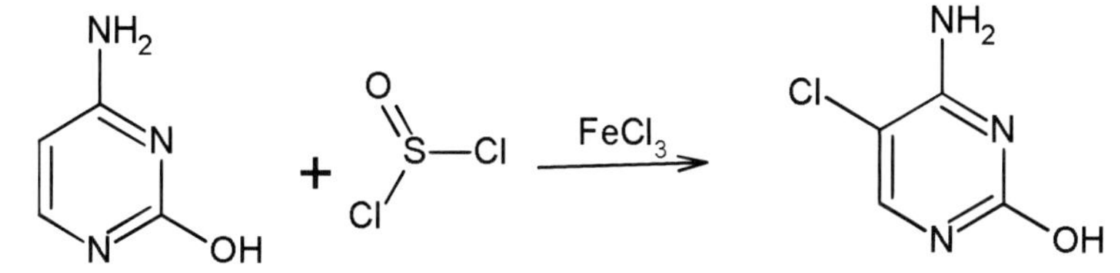
NH2
N
N
OH
+
O
S—Cl
Cl
FeCl3
Cl
NH2
N
N
OH

4-amino-2-hydroxypyrimidine

4-amino-5-chloro-2-hydroxypyrimidine

**Figure 23.** Reaction of 4-amino-2-hydroxypyrimidine with sulfuryl chloride.

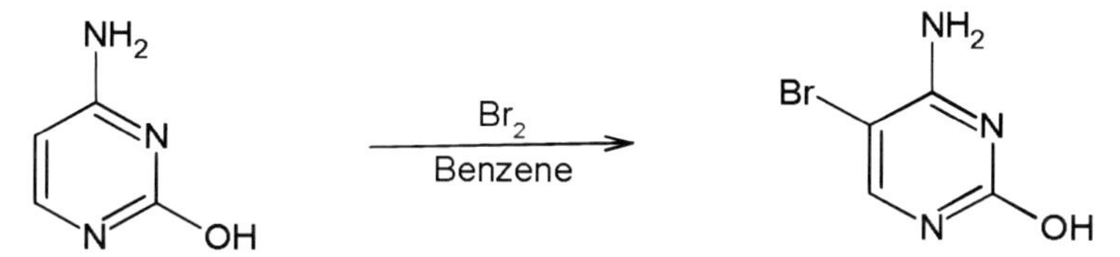
NH2
N
N
OH
Br2
Benzene
Br
NH2
N
N
OH

4-amino-2-hydroxypyrimidine

4-amino-2,5-bromo-2-hydroxypyrimidine

**Figure 24.** Reaction of 4-amino-2-hydroxypyrimidine with bromine.

## 5.6. Diazo Coupling Reaction of Pyrimidine

In the diazo reaction of pyrimidine, pyrimidine and a diazo reagent combine to generate a diazo compound. Organic compounds known as diazo compounds contain the diazonium functional group(-N=N-). To perform the diazo reaction with a pyrimidine, pyrimidine is commonly treated with a diazotizing reagent, such as sodium nitrite (NaNO2), in the presence of an acid, such as hydrochloric acid (HCl). A pyrimidine diazonium salt is produced as the reaction progresses through the diazotization of pyrimidine. Due to the diazonium electrophile's relative weakness, effective reactions require strongly nucleophilic reactants. In the case of pyrimidines, effective coupling at the C5 position often necessitates the presence of two strong electron-releasing substituents at either the C2 and C4 positions or the C6 position. For instance, a desired product, 4-amino-5-(phenyldiazenyl) pyrimidine-2-ol, is produced in a good yield when 4-amino-2-hydroxy pyrimidine interacts with a diazonium salt (Figure 25) (Brown, D.J., R.F. Evans, 1994).

## 5.7. Vilsmeier-Haack Reaction

The Vilsmeier-Haack reaction is used to introduce a formyl group (CHO) at specific positions on the pyrimidine ring. Usually, to carry out the reaction, pyrimidine is treated with a forming agent like a chloroformate (like oxalyl chloride) or a formamide derivative (like N, N-dimethylformamide) and an acid catalyst like phosphorus oxychloride or anhydrous aluminium chloride. The Vilsmeier-Haack reaction proceeds via the formation of an iminium intermediate, which is then followed by subsequent hydrolysis or reduction to produce the desired aldehyde or ketone product. The functionalization of pyrimidines using this technique is widely utilised in organic synthesis and has uses in agrochemical and pharmaceutical research (K. Moriyama, 1989).

Under Vilsmeier-Haack conditions, when the compound was combined with phosphorus oxychloride (POCl3) and N, N-dimethylformamide (DMF) resulted in the formation of 6-formyl-1,3-dimethylthieno[2,3-d]pyrimidine-2,4(1H,3H)-dione (Figure 26) (Hirota et al., 1990).

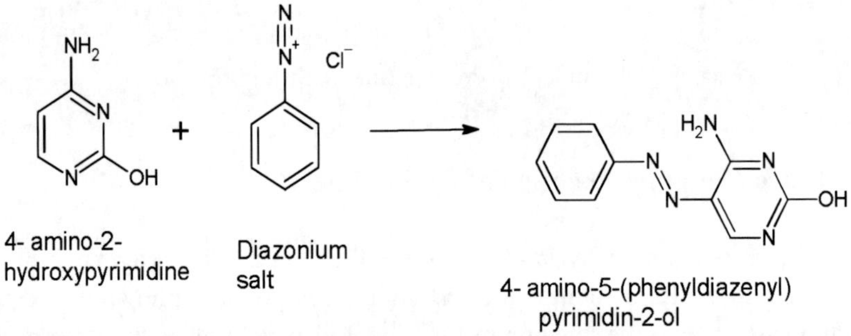
4- amino-2-hydroxypyrimidine
Diazonium salt
4- amino-5-(phenyldiazenyl) pyrimidin-2-ol

**Figure 25.** Reaction of 4-amino-2-hydroxypyrimidine with diazonium salt.

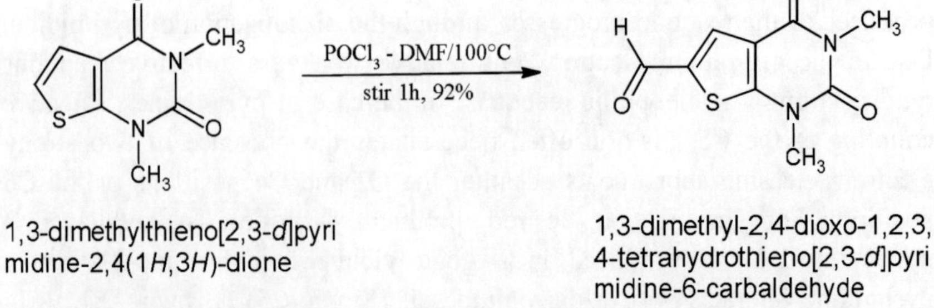
POCl$_3$ - DMF/100°C
stir 1h, 92%
1,3-dimethylthieno[2,3-*d*]pyrimidine-2,4(1*H*,3*H*)-dione
1,3-dimethyl-2,4-dioxo-1,2,3,4-tetrahydrothieno[2,3-*d*]pyrimidine-6-carbaldehyde

**Figure 26.** Reaction of phosphorus oxychloride and N, N-dimethylformamide.

# 6. Medicinal Uses

## 6.1. Antibacterial Activity

Antibacterial agents are medicines that combat bacteria. They include antifolates and sulfa drugs.

### *6.1.1. Antifolates are Medications that Work against Folic Acid*

They are composed of 2-amino-4-hydroxypyrimidines and 2,4-diaminopyrimidines, which are inhibitors of dihydrofolate reductase (DHFR). Some notable antifolates are Brodimoprim, Iclaprim, and Trimethoprim, which are effective against different types of bacteria. Pyrimethamine is another type of antifolate that works against malaria by selectively inhibiting DHFR in malarial plasmodia (Figure 27) (Sharma et al., 2014).

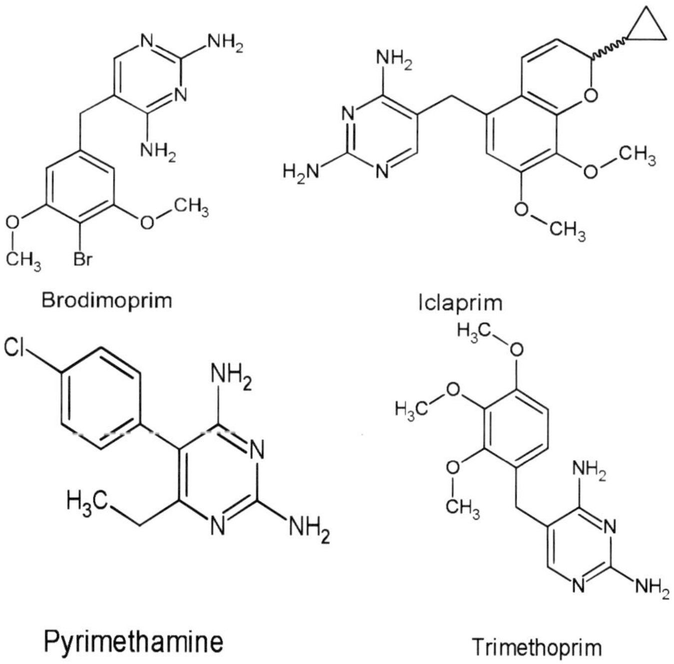
Brodimoprim
Iclaprim
Pyrimethamine
Trimethoprim

**Figure 27.** Antimicrobial drugs.

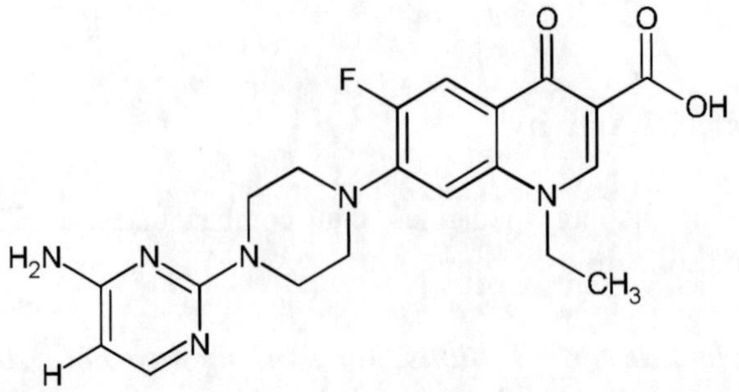
F
O
O
OH
N
N
CH3
H2N
N
N
N
H

**Figure 28.** Pyrimidine – conjugated flouroquinolone.

### *6.1.2. A Class of Medication Known as Sulfa Medicines Includes Pyrimidine*

The pyrimidine molecule undergoes a variety of chemical modifications that determine the different types of sulfa medicines. Sulfadiazine, Sulfamerazine, and Sulfadimidine are mono- and disubstituted sulfa medications that are frequently used in patients with acute bacterial infections, cerebrospinal meningitis, and penicillin allergies. Trisubstituted sulfa medications, such as sulfadoxine and sulfisomidine, are used as a combination of sulfa therapy and as a prophylactic against malaria. Trimethoprim and sulfa medications are frequently combined to create potent antibacterial medications (Sharma et al., 2014).

### *6.1.3. New Advances in Pyrimidine as an Antibacterial Drug (Pyrimidine-Conjugated Fluoroquinolones)*

Development of pyrimidine-conjugated fluoroquinolones to address the dreaded drug reaction. Most of the target pyrimidine derivatives can inhibit the growth of tested bacteria, especially the 4-aminopyrimidine compounds showed broad-spectrum antibacterial and low cytotoxicity and good antibacterial activity against *Enterococcus faecalis* and resistance to the MIC of norfloxacin and ciprofloxacin. As little as 0.25 μg/mL. Active drugs containing bactericidal bacteria rapidly inhibited biofilm formation and were less likely to develop resistance than norfloxacin and ciprofloxacin. Further research has shown that this compound promotes the accumulation of ROS in bacterial cells and interacts with DNA to form DNA complexes that promote bacterial death. ADME analysis showed that the compound had similar chemical and pharmacokinetic properties. These results suggest that pyrimidine-conjugated fluoroquinolones are promising potential antiviral agents worthy of further investigation (Figure 28) (Tan et al., 2022).

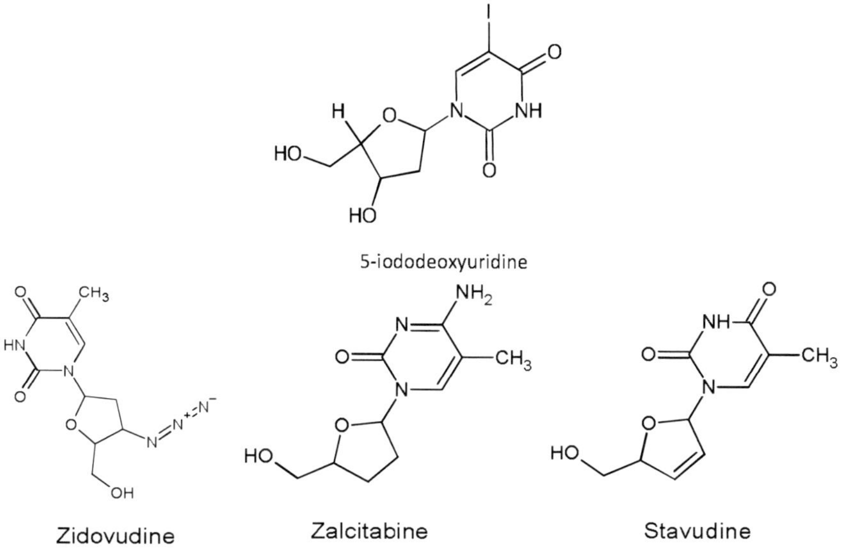
5-iododeoxyuridine
Zidovudine
Zalcitabine
Stavudine

**Figure 29.** Antiviral agents.

## 6.2. Antiviral Activity

Pyrimidine derivatives have good antiviral properties, with 5-iododeoxyuridine and IDU (5-iodo-2'- deoxyuridine) being commonly used for viral infections. 5-Trifluoromethyl-2-deoxyuridine is effective against infections resistant to IDU therapy. 1-(3-Azido-2,3-dideoxypentofuranosyl)-5-methyl 2,4(1H,3H) -pyrimidinedione inhibits HIV replication and has been approved for use against AIDS and AIDS-related complex (ARC). Lamivudine is effective against AIDS when used with zidovudine, while cidofovir is used to treat cytomegalovirus (CMV). Zidovudine is active against retroviruses and is used for AIDS and T-cell leukemia. Zalcitabine and stavudine, combined with zidovudine, show significant activity against HIV (Figure 29) (Sharma et al., 2014), (Siddiqui et al., 2007), (Krečmerová et al., 2017).

*Pyrimidine-based drugs for SARS CoV-2.*

Approximately 18,000 drugs with antiviral activity using live bacteria from the human respiratory tract and confirmed 122 drugs with antiviral activity and SARS-CoV-2 selection. Among these drug candidates are 16 nucleoside analogs, the largest class of antibiotics in clinical use. This includes the antiviral drugs remdesivir and molnupiravir approved for COVID-19.

RNA viruses rely on a host-supplied source of nucleoside triphosphate to be effective, a group of host nucleoside biosynthesis inhibitors is identified as potential antiviral agents. In addition, the combination of pyrimidine biosynthesis inhibitors with antimicrobial nucleoside analogs synergistically inhibted SARS-COV-2 infection (Figure 30) (Schultz et al., 2022).

## 6.3. Antifungal Activity

Pyrimidines have antifungal properties and are utilized as nucleoside antifungal agents. Flucytosine, a fluorinated pyrimidine, is effective against serious systemic infections caused by susceptible strains of Candida and Cryptococcus. Hexitidine is a medication mainly prescribed for the treatment of aphthous ulceration (Figure 31).

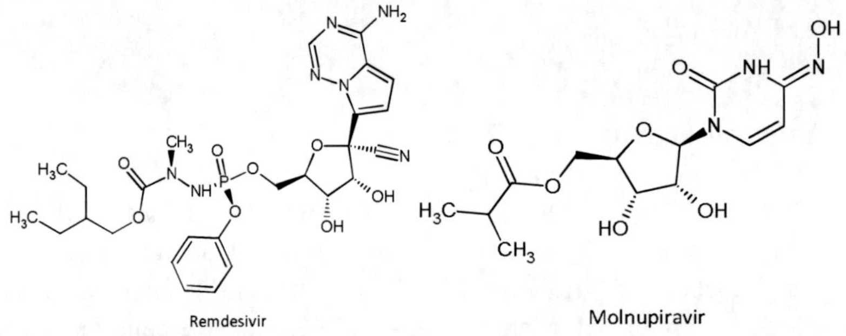
Remdesivir
Molnupiravir

**Figure 30.** Pyrimidine based drugs for SARS Cov-2.

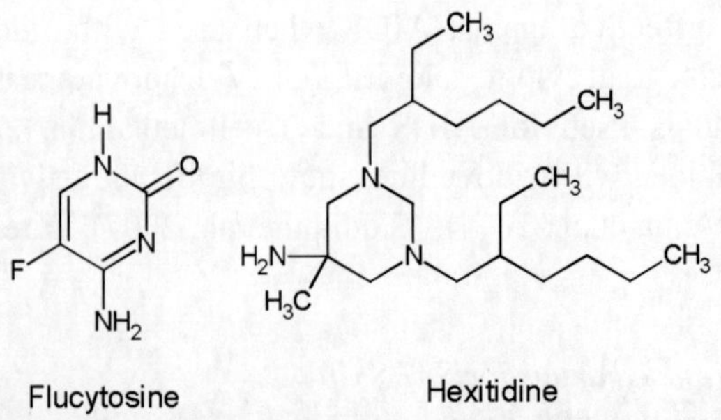
Flucytosine
Hexitidine

**Figure 31.** Antifungal agents.

Ishak et al. synthesized pyrazole-[1,5-a] pyrimidine and pyrimidine derivatives and examined them for their antimicrobial activities (Figure 32). A new group of pyrazole-[1,5-a] pyrimidines were also screened for antifungal activity (*Candida albicans, Aspergillus fumigatus, Geotrichum candida, Cephalosporium racemosa*). 5,7-di(furan-2-yl)-3-(p-tolidiazenyl) pyrazolo[1,5-a] pyrimidin-2-amine, showed good activity than (thiophen-2-yl)-3-(p-tolyldiazenyl) pyrazolo[1,5-a] pyrimidin-2-amine.

All structures of the novel compounds were supported by IR, 1H-NMR, 13C-NMR, GC-MS, and CHN analysis. *In vitro*, the antifungal activity of newly synthesized pyrimidine derivatives against *Candida albicans, Penicillium*, and *Aspergillus niger* was evaluated by the agar plate diffusion method with amphotericin B as the standard drug (Figure 33) (Andrews & Ahmed, 2014), (Ibrahim Wahbi et al., 2013).

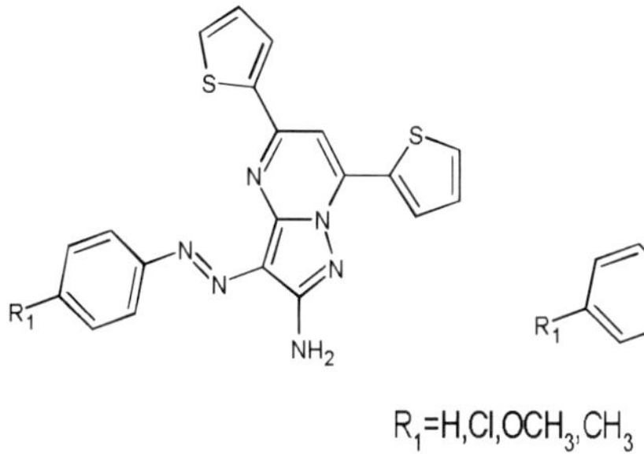
R1=H,Cl,OCH3,CH3

**Figure 32.** Pyrimidine derivatives.

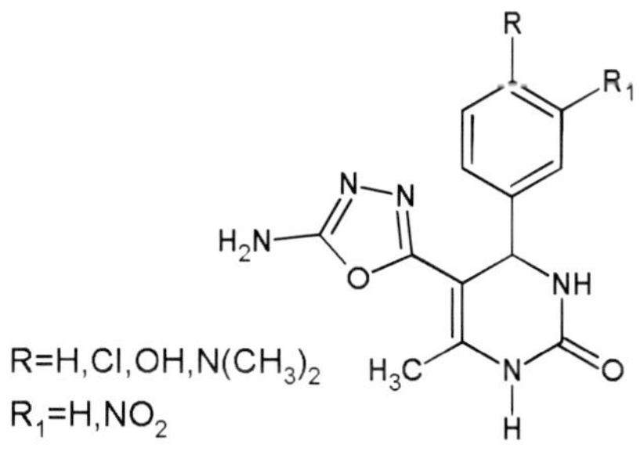
R=H,Cl,OH,N(CH3)2
R1=H,NO2

**Figure 33.** 1,3,4-oxadiazole derivatves.

## 6.4. Anti-Alzheimer's Activity

Alzheimer's disease (AD) is a neurodegenerative condition that results in memory loss and cognitive impairment. Extracellular amyloid-(A) plaques, intracellular neurofibrillary tangles made of hyperphosphorylated tau, and a persistent loss of neurons are the hallmarks of AD brain disease. With the mitochondrial respiratory chain interacting with the de novo pyrimidine biosynthesis pathway, respiratory complex IV function deficit is a typical symptom of late-onset AD. Pesini et al. claim that individuals with AD may have reduced pyrimidine nucleotide levels, which can have an impact on a variety of cellular processes, including the development of neuronal membranes and synapse formation. Neuronal differentiation is reduced when oxidative phosphorylation activity is inhibited, however uridine treatment can improve neuronal differentiation by connecting these pathways to pyrimidine nucleotides. Alzheimer's patients' brains were discovered to have altered mRNA levels for genes involved in both the pyrimidine salvage pathway and de novo pyrimidine production (Nerkar, 2021), (Aamir et al., 2023).

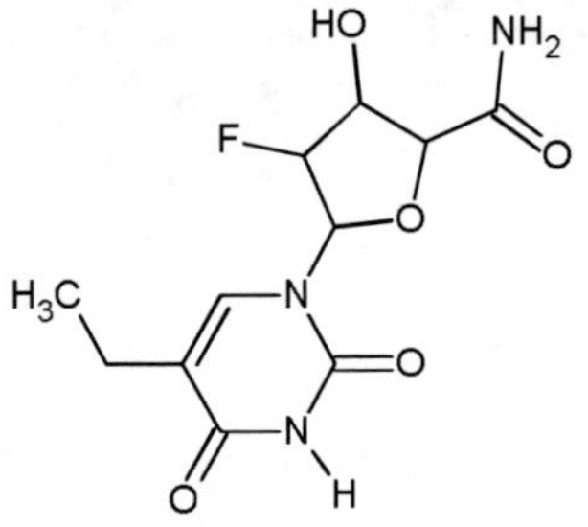

4'- carboxamide pyrimidine nucleoside

**Figure 34.** Antihepatic drugs.

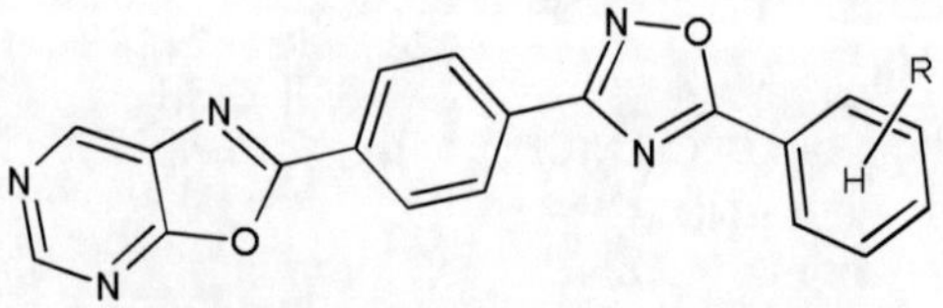

1,2,4-oxadiazole linked 4-oxazolo[5,4-d] pyrimidine

**Figure 35.** Anticancer drugs.

## 6.5. Anti-Hepatic Activity

Hepatitis B virus and hepatitis C virus are ordinary liver infections, they can co-exist. Along with it, there is a high risk of developing liver cancer, hepatocellular carcinoma, and cirrhosis, all of which can be fatal. A new class of pyrimidine nucleosides with a 4-carboxymethyl and 4-carboxamide functional group was synthesized and evaluated for their anti-HCV activity (Figure 34). The anti-HCV properties of these compounds were found to be comparable to those of ribavirin (Nerkar, 2021).

## 6.6. Anticancer Activity

Cancer is one of the life-threatening diseases. MTT assays were employed to evaluate the anticancer potential of a range of pyrimidine-bridged combretastatin derivatives against MCF-7 (breast cancer) and A549 (lung cancer) cell lines. The MTT assay is a colorimetric method utilized to quantify cell proliferation. Significantly, the majority of the synthesized compounds demonstrated robust anticancer activity, exhibiting IC50 values within the low micromolar range. Four human cancer cell lines breast cancer, lung cancer, ovarian cancer, and colon cancer can be treated by an arrangement of 1,2,4-oxadiazole connected 4-oxazolo[5,4-d]pyrimidine derivatives (Figure 35) (Nerkar, 2021), (Perupogu et al., 2020), (Kumar et al., 2018).

*Pyrimidine and fused pyrimidine derivatives have protein kinase inhibitors for the treatment of cancer.*

Pyrimidine and its fused derivatives, such as pyrazolo[3,4-d] pyrimidine, pyrido[2,3-d] pyrimidine, quinazoline, and furo[2,3-d]pyrimidine compounds, have gained significant attention for their potential as protein kinase inhibitors in cancer treatment. These compounds exhibit diverse biological properties and are considered bioisosteres with purines. Several pyrimidine derivatives, including pyrazolo[3,4-d] pyrimidine, pyrido[2,3-d] pyrimidine, quinazoline, and furo[2,3-d] pyrimidine, have demonstrated promising anticancer activity. The mechanisms of action for these pyrimidine derivatives involve the inhibition of protein kinases, which are crucial enzymes regulating cell growth, differentiation, migration, and metabolism. This review focuses on the anticancer implications of select pyrimidine and fused pyrimidine derivatives, specifically their selective inhibition of protein kinases. The review also explores structure-activity relationships and synthetic pathways utilized for the construction of these scaffolds, aiming to aid medicinal chemists in developing novel pyrimidine compounds with enhanced selectivity as anticancer agents (Abdellatif & Bakr, 2021), (Shu et al., 2020).

### 6.7. Anti-Thyroid Activity

The effective drugs used against hyperthyroidism are 2-Thiouracil and its alkyl analogs, phenobarbital, and propylthiouracil (Figure 36) (K. S. Jain, T. S. Chitre, P. B. Miniyar, M. K. Kathiravan, V. S. Bendre, V. S. Veer, 2015).

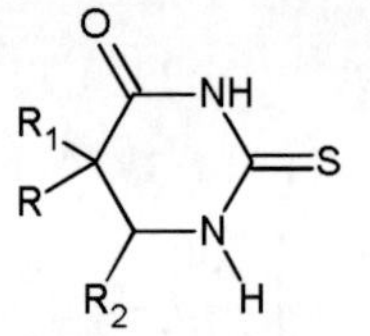

$R=R_1=R_2=H$; 2-thiouracil
$R=R_1=H, R_2=C_3H_7$; propylthiouracil
$R=R_1=C_2H_5, R_2=O$; thiobarbital

**Figure 36.** Antithyroid drugs.

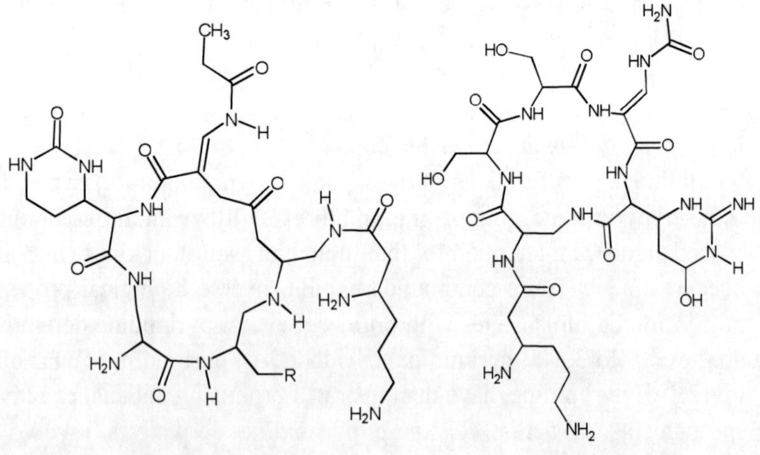

Capreomycin Viomycin

**Figure 37.** Antitubercular drugs.

### 6.8. Antitubercular Activity

Pyrimidine derivatives capreomycin, derived from *Streptomyces Capreolus* belongs to the second-line category of antitubercular drug. Viomycin use in experimental tuberculosis (Figure 37) (Tolba et al., 2022), (Chandrashekaraiah et al., 2014).

### 6.9. Anthelmintics Activity

Against ringworms and pinworms, Pyrantel pamoate is utilized. It induces spastic paralysis by depolarising neuromuscular blocking agents (Figure 38) (Patil, 2023).

### 6.10. Anti-Inflammatory Activity

Thienopyrimidines derivatives are effective against inflammation. Thiadiazole containing phenyl-(5-(thieno[2,3-d] pyrimidin-4-yloxymethyl)-(1,3,4)thiadiazol-2-yl)-amine was found to be most effective. A series of novel pyrazolo[3,4-d] pyrimidines were synthesized and evaluated to construct a new scaffold between 4-amino pyrazole (3,4-d) pyrimidine and different 4-substituted benzoyl chlorides in N, N-dimethyl formamide solvent (Figure 39). These compounds were evaluated for protective activity in the carrageenan edema model. All of these compounds were characterized by physical and spectroscopic data. Chemicals modified with electron-withdrawing groups, such as chlorine 6b and fluorine 6d, showed significant protective effects at 3 mg/kg compared with the standard indomethacin drug at 10 mg/kg and, respectively.

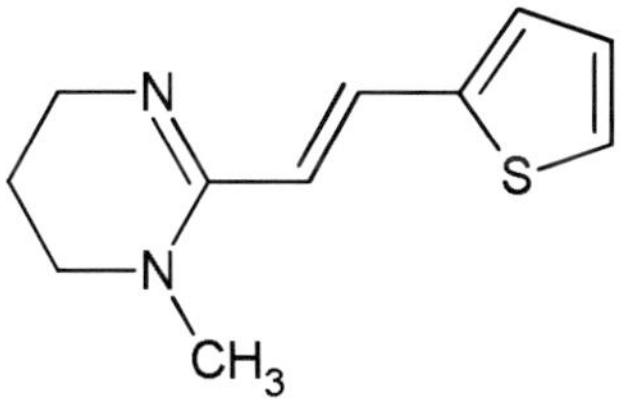
N
S
N
CH3

Pyrantel pamoate

**Figure 38.** Anthelmintics drugs.

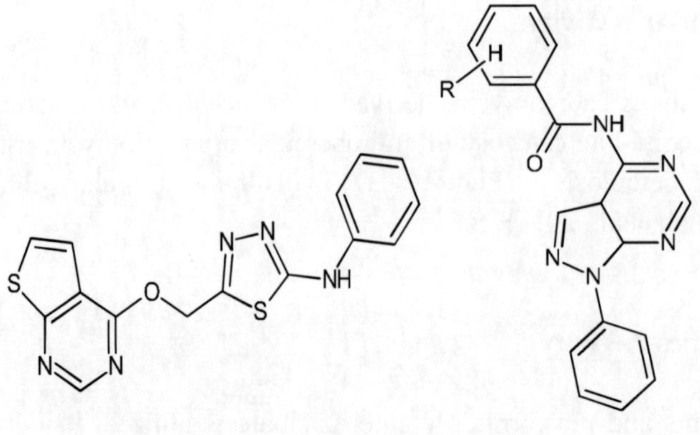
R
H
NH
O
N
N
N
N
N
N–N
NH
S
O
S
N
N
Thienopyrimidines derivatives
Pyrazolo [3,4-d] pyrimidines

**Figure 39.** Anti-inflammatory drugs.

Kumaresan et al. reported the synthesis of some new pyrimidine 2-amines and their anti-inflammatory effects on locally induced edema in Balb/c mice. Compounds were established by infrared, 1H, 13C NMR, and mass spectrometry techniques. 4-(9H-Fluoren-2-yl)-6-phenylpyrimidine 2-amine and 4-(4-[diphenyl amino]phenyl) pyrimidine-2-amine 4,6-Bis-(9H-fluorene-2)-yl) pyrimidine-2-amine and 3-(3-[9H-Fluoren-2-yl]-3-oxoprop-1en-1-yl)-4H-chromene-4-one (Dansena et al., 2015), (Ashour et al., 2013).

## 7. Latest Pyrimidine Studies

### 7.1. Effective Approaches for Identifying Chromosomal Markers in *Lactococcus Lactis*: Analyzing Mutants of the Pyrimidine Salvage Pathway Obtained through Positive Selections

The goal of the study was to create efficient methods for recognising chromosomal markers in Lactococcus lactis, a common bacterium utilised in biotechnology and dairy fermentations. The study concentrated on the analysis of positive selection mutants of the pyrimidine salvage pathway in L. lactis. The pyrimidine salvage process is vital for producing pyrimidine nucleotides, which are necessary building blocks for DNA and RNA. Bacterial growth or survival may be hampered by mutations in this mechanism. Finding mutants with abnormalities in this pathway is therefore crucial for comprehending the underlying genetic mechanisms and their possible uses.

To do this, the scientists used a positive selection technique that made it possible to isolate mutants with obstructed pyrimidine salvage routes. Combining chemical mutagenesis with a growth medium devoid of exogenous pyrimidines was their method. This method put pressure on mutants to grow and survive even in the absence of pyrimidine nucleotides. The researchers produced a group of mutants from L. lactis by mutagenesis and positive selection that may have abnormalities in the pyrimidine salvage pathway. The discovered mutations were subsequently confirmed and described using a variety of molecular techniques, including PCR, DNA sequencing, and gene expression studies.

The results of the study demonstrated the successful isolation of *L. lactis* mutants with disruptions in the pyrimidine salvage pathway. Through genetic analysis, the researchers identified specific chromosomal markers associated with the mutations. They also assessed the growth characteristics and metabolic profiles of the mutants to gain insights into the consequences of the disrupted pathway.

The findings of this study contribute to the understanding of the pyrimidine salvage pathway in *L. lactis* and provide valuable genetic markers for future research and applications. The identified mutants could be further studied to elucidate the specific genetic mechanisms involved and explore their potential applications in dairy fermentation and biotechnology.(Martinussen & Hammer, 1995)(Nilsson & Lauridsen, 1992)

## 7.2. A Novel Pyrimidine-Based Red Fluorescent Probe Model for Detection of Hydrogen Peroxide in Parkinson's Disease

Parkinson's disease (PD) is a neurodegenerative disorder characterized by the progressive loss of dopaminergic neurons in the brain. Oxidative stress, which involves an imbalance between the production of reactive oxygen species (ROS) and the antioxidant defense system, is believed to play a crucial role in the development and progression of PD. Hydrogen peroxide ($H_2O_2$) is a major ROS that is involved in oxidative damage.

In recent years, the development of fluorescent probes for the detection of $H_2O_2$ has gained significant attention as a valuable tool for studying oxidative stress-related diseases, including PD. In this context, a group of researchers has proposed a novel pyrimidine-based red fluorescent probe model specifically designed for the detection of $H_2O_2$ in Parkinson's disease. The research team synthesized and characterized the pyrimidine-based probe, which exhibited excellent sensitivity and selectivity towards $H_2O_2$. The probe was designed to undergo a red-shifted fluorescence emission upon reaction with $H_2O_2$, enabling easy visualization and quantification of $H_2O_2$ levels in biological samples.

The researchers validated the probe's performance in both cell cultures and animal models of PD. They observed a significant increase in $H_2O_2$ levels in PD-related cellular models and PD mouse brain tissues compared to control samples. Furthermore, they demonstrated that the probe selectively detected $H_2O_2$ over other ROS and reactive nitrogen species (RNS), highlighting its specificity. Additionally, the pyrimidine-based probe model showed good stability, low cytotoxicity, and satisfactory photostability, making it a reliable tool for long-term monitoring of $H_2O_2$ levels in PD research. Its successful application in detecting $H_2O_2$in PD models opens up new possibilities for understanding the oxidative stress mechanisms involved in PD pathogenesis and for evaluating potential therapeutic interventions targeting ROS.

Overall, the development of this novel pyrimidine-based red fluorescent probe model for $H_2O_2$ detection in Parkinson's disease provides a valuable tool for researchers to study the role of oxidative stress and explore new therapeutic strategies in combating PD.(Qiu et al., 2019),(Li et al., 2018)

## 7.3. Inhibition of Copper Corrosion in Synthetic Seawater Using Pyrazole Pyrimidine Derivative

The study investigated the potential of a pyrazole pyrimidine derivative as a corrosion inhibitor for copper in synthetic seawater. Corrosion of copper in marine environments is a significant concern due to its extensive use in marine structures and equipment. The researchers aimed to evaluate the inhibitory properties of the synthesized compound and its effectiveness in preventing copper corrosion.

The pyrazole pyrimidine derivative was synthesized and characterized using various spectroscopic techniques. The researchers conducted corrosion inhibition experiments by immersing copper samples in synthetic seawater containing different concentrations of the inhibitor. They employed electrochemical techniques, such as potentiodynamic polarization and electrochemical impedance spectroscopy, to assess the corrosion behavior of copper in the presence and absence of the inhibitor.

The results showed that the pyrazole pyrimidine derivative exhibited excellent corrosion inhibition properties for copper in synthetic seawater. At higher inhibitor concentrations, a significant reduction in the corrosion rate of copper was observed. The inhibitor formed a protective film on the copper surface, effectively hindering the corrosion process.

The inhibition efficiency was found to increase with increasing inhibitor concentration and immersion time. The synthesized compound demonstrated a high degree of stability and adherence to the copper surface, ensuring long-term corrosion protection. The researchers also investigated the influence of temperature and pH on the inhibitor's performance and observed favorable inhibition effects under a wide range of conditions.

The study highlighted the potential of the pyrazole pyrimidine derivative as an effective corrosion inhibitor for copper in synthetic seawater. The compound's ability to form a protective film on the metal surface and its stability make it a promising candidate for corrosion prevention in marine environments. Further research and practical applications of the inhibitor can contribute to the development of more efficient corrosion protection strategies for copper-based marine structures and equipment (Echihi et al., 2020), (Ansari et al., 2015).

## 7.4. Novel Pyrimidine-Tagged Silver Nanoparticles-Based Fluorescent Immunoassay for the Detection of *Pseudomonas aeruginosa*

Pseudomonas aeruginosa is a common pathogenic bacterium responsible for severe infections, particularly in hospital settings. Rapid and accurate detection of *P. aeruginosa* is crucial for effective disease management and prevention of its spread. In this study, researchers developed a novel fluorescent immunoassay using pyrimidine-tagged silver nanoparticles (AgNPs) for the specific detection of *P. aeruginosa.*

The researchers synthesized pyrimidine-tagged AgNPs and functionalized them with specific antibodies targeting *P. aeruginosa*. The binding of the antibodies to the bacterial surface facilitated the formation of antibody-antigen complexes, leading to the aggregation of the AgNPs. This aggregation resulted in a significant change in the fluorescence properties of the nanoparticles, allowing for easy detection and quantification.

The developed immunoassay demonstrated excellent specificity towards *P. aeruginosa*, with negligible cross-reactivity towards other common bacterial species. The sensitivity of the assay was evaluated using various concentrations of *P. aeruginosa*, and a linear relationship between the fluorescence intensity and bacterial concentration was observed. The limit of detection achieved by the assay was notably low, indicating its potential for early detection of *P. aeruginosa* infections.

Furthermore, the immunoassay was successfully applied to the analysis of clinical samples, including patient sputum and urine samples. The results obtained from the assay were compared to traditional culture-based methods, and a high degree of agreement was observed, confirming the reliability and accuracy of the developed fluorescent immunoassay.

The advantages of this novel approach include its simplicity, rapidity, and high sensitivity, making it a promising tool for the early and specific detection of *P. aeruginosa* infections. The use of pyrimidine-tagged AgNPs provides a unique and robust platform for fluorescent immunoassays, offering potential applications in clinical diagnostics and monitoring of *P. aeruginosa*-related infections in healthcare settings.

In conclusion, the development of this novel pyrimidine-tagged AgNPs-based fluorescent immunoassay represents a significant advancement in the detection of *P. aeruginosa*. The assay's specificity, sensitivity, and compatibility with clinical samples make it a promising diagnostic tool with the potential to improve patient outcomes through early and accurate detection of *P. aeruginosa* infections (Ellairaja et al., 2017), (Kaur et al., 2015).

## 7.5. Bioinspired Pyrimidine-Containing Cationic Polymers as Efficient Nano-Carriers for DNA and Protein Delivery

The efficient delivery of DNA and proteins into cells is a crucial requirement in various biomedical applications, including gene therapy and protein-based therapeutics. In this study, researchers developed bioinspired pyrimidine-containing cationic polymers as highly efficient nano-carriers for DNA and protein delivery.

The researchers designed and synthesized cationic polymers inspired by the structure of natural biomolecules, specifically incorporating pyrimidine-containing moieties. These polymers exhibited excellent biocompatibility and demonstrated high binding affinity towards DNA and proteins. The unique chemical structure of the polymers allowed for efficient complexation and protection of the cargo molecules.

The researchers evaluated the delivery performance of the pyrimidine-containing cationic polymers using *in vitro* experiments. They demonstrated successful encapsulation of DNA and proteins within the polymer nanoparticles and observed efficient cellular uptake of the cargo by different cell types. The nano-carriers displayed minimal cytotoxicity and maintained the stability of the cargo molecules during the delivery process.

The study also investigated the ability of the polymer nano-carriers to facilitate intracellular delivery and transfection of DNA. The results showed significantly enhanced gene expression compared to conventional non-viral delivery systems. Moreover, the researchers demonstrated the successful delivery of proteins using the polymer nano-carriers, leading to functional protein expression in the target cells.

The bioinspired pyrimidine-containing cationic polymers offered several advantages as nano-carriers for DNA and protein delivery. Their structural similarity to natural biomolecules improved their biocompatibility and minimized potential immune responses. The polymers also exhibited superior stability and protection of the cargo, enabling efficient delivery and subsequent biological activity of DNA and proteins.

Overall, this study highlights the potential of bioinspired pyrimidine-containing cationic polymers as highly efficient nano-carriers for DNA and protein delivery. Their ability to encapsulate and protect cargo molecules, promote cellular uptake, and facilitate successful intracellular delivery makes them promising candidates for various biomedical applications. Further research and optimization of these polymer nano-carriers could contribute to the development of improved gene therapy and protein-based therapeutics (Manuscript, 2020).

## 7.6. Gold and Platinum Nanoparticles Capped with Novel Pyrimidine-Based Ligands

Gold (Au) and platinum (Pt) nanoparticles have gained significant attention in various fields, including catalysis, sensing, and biomedical applications, due to their unique properties and reactivity. In this study, researchers developed and characterized gold and platinum nanoparticles capped with novel pyrimidine-based ligands, exploring their potential applications.

The researchers synthesized the pyrimidine-based ligands and employed them to cap the surfaces of gold and platinum nanoparticles. The ligands exhibited a strong affinity for the metal surfaces, providing stable and well-dispersed nanoparticles. The unique chemical properties of the ligands allowed for precise control over the nanoparticle size and surface properties.

The synthesized gold and platinum nanoparticles capped with the pyrimidine-based ligands demonstrated enhanced stability and dispersibility in various solvents. The ligand shell also imparted improved solubility and compatibility, facilitating the functionalization and surface modification of the nanoparticles. The researchers further characterized the nanoparticles using techniques such as transmission electron microscopy (TEM), X-ray diffraction (XRD), and spectroscopic methods. The study also explored the catalytic applications of the pyrimidine-capped nanoparticles. The ligands' chemical nature and their interaction with the metal surfaces provided unique catalytic properties to the nanoparticles. The researchers observed enhanced catalytic activities for various reactions, including oxidation and reduction processes, making these nanoparticles potential catalysts for diverse chemical transformations. Additionally, the nanoparticles capped with the pyrimidine-based ligands exhibited intriguing optical properties, including plasmonic behavior in the case of gold nanoparticles. This optical activity opens up possibilities for their application in sensing, imaging, and photonic devices.

In summary, the development of gold and platinum nanoparticles capped with novel pyrimidine-based ligands offers new opportunities for various applications. The ligand-mediated stability, solubility, and catalytic properties of the nanoparticles make them promising candidates for catalysis, sensing, and optical applications. The ability to precisely control the nanoparticle properties through ligand design and modification provides a versatile platform for tailoring and optimizing their performance in different fields. Further research and exploration of these pyrimidine-based ligand-capped nanoparticles could lead to advancements in nanotechnology and materials science (Vadivel et al., 2017).

## Conclusion

The review article on pyrimidine concludes by giving a thorough description of its chemistry, synthesis processes, reactions, medical applications, and most recent research. Various synthetic techniques are covered in the article, including Pinner-type synthesis, sequential Suzuki coupling/condensation reactions, green one-pot synthesis from nitriles, oxidative copper-catalyzed condensation, Kemp's three- and four-component syntheses, Kirchner three-component synthesis, and Diels-Alder cycloaddition. Additionally, it covers microwave-based pyrimidine synthesis techniques and reactions, such as Vilsmeier-Haack and diazo coupling as well as alkylation, oxidation, reduction, nitration, and halogenation. Also included are the therapeutic applications of pyrimidine, with a focus on its antibacterial, antiviral, antifungal, anti- Alzheimer's, antidiabetic, anti-hepatic, anticancer, anti-thyroid, antitubercular, anthelmintic, and anti-inflammatory properties. The review article discusses the most recent findings on pyrimidine, such as its use in the identification of chromosomal markers in Lactococcus lactis, its use as a red fluorescent probe for hydrogen peroxide detection in Parkinson's disease, its capacity to stop copper corrosion in synthetic seawater, and its use in the development of novel fluorescent immunoassays and nano-carriers for DNA and protein delivery. The thorough examination of numerous synthesis techniques, varied chemical processes, and a wide range of therapeutic uses highlights the adaptability and importance of pyrimidine in both theoretical and real-world settings.

## References

Aamir, M., Saeed, M., Shbeer, A. M., Al-ghorbani, M., Rauf, A., Wilairatana, P., Mannan, A., Sadiq, A., Farooq, U., and Rashid, U. (2023). Biomedicine & Pharmacotherapy Evaluation of pyrimidine/pyrrolidine-sertraline based hybrids as computational studies. *Biomedicine & Pharmacotherapy*, 159(December 2022), 114239. https://doi.org/10.1016/j.biopha.2023.114239.

Abdelghani, E., Said, S. A., Assy, M. G., and Abdel Hamid, A. M. (2017). Synthesis and antimicrobial evaluation of some new pyrimidines and condensed pyrimidines. *Arabian Journal of Chemistry*, 10, S2926–S2933. https://doi.org/10.1016/j.arabjc.2013.11.025.

Abdellatif, K. R. A., and Bakr, R. B. (2021). Pyrimidine and fused pyrimidine derivatives as promising protein kinase inhibitors for cancer treatment. *Medicinal Chemistry Research*, 30(1), 31–49. https://doi.org/10.1007/s00044-020-02656-8.

Andrews, B., and Ahmed, M. (2014). *International Journal of Chemical Studies Synthesis and characterization of pyrimidine bearing 1, 3, 4- oxadiazole derivatives and their potential antifungal action*. 1(4), 32–39.

Ansari, K. R., Sudheer, Singh, A., and Quraishi, M. A. (2015). Some Pyrimidine Derivatives as Corrosion Inhibitor for Mild Steel in Hydrochloric Acid. *Journal of Dispersion Science and Technology*, 36(7), 908–917. https://doi.org/10.1080/01932691.2014.938349.

Antonello, A., Hrelia, P., Leonardi, A., Marucci, G., Rosini, M., Tarozzi, A., Tumiatti, V., and Melchiorre, C. J. (2005). *Med. Chem.*, 28, 48.

Arcelo, M., Ravina, E., Masaguer, C. F., Dominguez, E., Areias, F. M., Brea, J., and Loza, M. I. (2007). *Bioorg. Med. Chem. Lett.*, 17, 4873.

Ashour, H. M., Shaaban, O. G., Rizk, O. H., and El-Ashmawy, I. M. (2013). Synthesis and biological evaluation of thieno [2′,3′:4,5] pyrimido[1,2-b][1,2,4]triazines and thieno[2,3-d][1,2,4]triazolo[1,5-a] pyrimidines as anti-inflammatory and analgesic agents. *European Journal of Medicinal Chemistry*, 62, 341–351. https://doi.org/10.1016/j.ejmech.2012.12.003.

Bagley, M. C., and Dale, J. W. B. (2002). *J. Chem. Commun.*, 1682.

Bagley, M. C., Hughes, D. D., and Taylor, P. H. (2003). Highly efficient synthesis of pyrimidines under microwave-assisted conditions. *Synlett*, 2, 259–261. https://doi.org/10.1055/s-2003-36781.

Boger, D. L. (1986). Diels-Alder Reactions of Heterocyclic Azadienes: Scope and Applications. *Chemical Reviews*, 86(5), 781–793. https://doi.org/10.1021/cr00075a004.

Brown, D.J., R.F. Evans, W. B. C. and M. D. F. (1994). *The Pyrimidine*. John Wiley and Sons, New York, USA., pp: 96-106.

Chandrashekaraiah, M., Lingappa, M., Deepu Channe Gowda, V., and Bhadregowda, D. G. (2014). Synthesis of some new pyrimidine-azitidinone analogues and their antioxidant, *in vitro* antimicrobial, and antitubercular activities. *Journal of Chemistry*, 2014. https://doi.org/10.1155/2014/847180.

Curphey, T. J., and Prasad, K. S. (1972). Diquaternary Salts. I. Preparation and Characterization of the Diquaternary Salts of Some Diazines and Diazoles. *Journal of Organic Chemistry*, 37(14), 2259–2266. https://doi.org/10.1021/jo00979a012

Dang, H. V., Le, H. T. B., Tran, L. T. B., Ha, H. Q., Le, H. V., and Phan, N. T. S. (2018). Copper-catalyzed one-pot domino reactions via C-H bond activation: Synthesis of 3-aroylquinolines from 2-aminobenzylalcohols and propiophenones under metal-organic framework catalysis. *RSC Advances*, 8(55), 31455–31464. https://doi.org/10.1039/c8ra05459b.

Dansena, H., Hj, D., and Chandrakar, K. (2015). Pharmacological potentials of pyrimidine derivative: A review. *Asian Journal of Pharmaceutical and Clinical Research*, 8(4), 171–177.

Deibl, N., Ament, K., Kempe, R., Deibl, N., Ament, K., and Kempe, R. (2015). A Sustainable Multi-Component Pyrimidine Synthesis_JACS 2015. *Journal of the American Chemical Society*, 137, 12804–12807.

Echihi, S., Benzbiria, N., Belghiti, M. E., El Fal, M., Boudalia, M., Essassi, E. M., Guenbour, A., Bellaouchou, A., Tabyaoui, M., and Azzi, M. (2020). Corrosion inhibition of copper by pyrazole pyrimidine derivative in synthetic seawater: Experimental and theoretical studies. *Materials Today: Proceedings*, 37(xxxx), 3958–3966. https://doi.org/10.1016/j.matpr.2020.09.264.

Ellairaja, S., Krithiga, N., Ponmariappan, S., and Vasantha, V. S. (2017). Novel Pyrimidine Tagged Silver Nanoparticle Based Fluorescent Immunoassay for the Detection of Pseudomonas aeruginosa. *Journal of Agricultural and Food Chemistry*, 65(8), 1802–1812. https://doi.org/10.1021/acs.jafc.6b04790.

Frutos, R. P., Wei, X., Patel, N. D., Tampone, T. G., Mulder, J. A., Busacca, C. A., and Senanayake, C. H. (2013). One-pot synthesis of 2,5-disubstituted pyrimidines from nitriles. *Journal of Organic Chemistry*, 78(11), 5800–5803. https://doi.org/10.1021/jo400720p.

Guirado, A., López-Caracena, L., López-Sánchez, J. I., Sandoval, J., Vera, M., Bautista, D., and Gálvez, J. (2016). A new, high-yield synthesis of 3-aryl-1,2,4-triazoles. *Tetrahedron*, 72(49), 8055–8060. https://doi.org/10.1016/j.tet.2016.10.045.

Hill, M. D., and Movassaghi, M. (2008). New strategies for the synthesis of pyrimidine derivatives. *Chemistry - A European Journal*, 14(23), 6836–6844. https://doi.org/10.1002/chem.200800014.

Hirota, K., Shirahashi, M., Senda, S., and Yogo, M. (1990). Pyrimidines. 65. Synthesis of 6-substituted thieno[2,3-d]pyrimidine-2,4(1H,3H)-diones. In *Journal of Heterocyclic Chemistry* (Vol. 27, Issue 3, pp. 717–721). https://doi.org/10.1002/jhet.5570270345.

Ibrahim Wahbi, H., Ishak, C. Y., and Metwally, N. H. (2013). Ishak et al. / Int. *J. Pharm. Phytopharmacol. Res*, 2(6), 407–411. www.eijppr.com.

In, S. (1931). Hückel's Rule Four Criteria for Aromaticity How Can You Tell Which Electrons are π Electrons? *Heterocyclic Aromatic Compounds*. 6–8.

Jiao, Y., Ho, S. L., and Cho, C. S. (2015). Copper-powder-catalyzed synthesis of pyrimidines from β-bromo α,β-unsaturated ketones and amidine hydrochlorides. *Synlett*, 26(8), 1081–1084. https://doi.org/10.1055/s-0034-1380410.

Jubeen, F., Iqbal, S. Z., Shafiq, N., Khan, M., Parveen, S., Iqbal, M., and Nazir, A. (2018). Eco-friendly synthesis of pyrimidines and its derivatives: A review on broad spectrum bioactive moiety with huge therapeutic profile. *Synthetic Communications*, 48(6), 601–625. https://doi.org/10.1080/00397911.2017.1408840.

K. Moriyama, T. N. and F. Y. (1989). *Heterocyclic Chem.*, 241.

Jain, K. S., Chitre, T. S., Miniyar, P. B., Kathiravan, M. K., Bendre, V. S., Veer, V. S., Shahane, S. R., and Shishoo, C. J. (2015). Biological and medicinal significance of pyrimidine. *European Journal of Medicinal Chemistry*, 97(1), 561–581. https://doi.org/10.1016/j.ejmech.2014.10.085.

Katritzky, A. R., Ramsden, C. A., Scriven,. E. F. V., and Taylor, R. J. K. (2008). *Comprehensive Heterocyclic Chemistry III.*, Eds. Pergamon: Oxford, U.K. 1–13.

Kaur, G., Raj, T., Kaur, N., and Singh, N. (2015). Pyrimidine-based functional fluorescent organic nanoparticle probe for detection of Pseudomonas aeruginosa. *Organic and Biomolecular Chemistry*, 13(16), 4673–4679. https://doi.org/10.1039/c5ob00206k.

Kidwai, M., and Mishra, A. D. E. O. (2004). *Using Inorganic Solid Supports*. 69(4), 247–253.

Krečmerová, M., Dračínský, M., Snoeck, R., Balzarini, J., Pomeisl, K., and Andrei, G. (2017). New prodrugs of two pyrimidine acyclic nucleoside phosphonates: Synthesis and antiviral activity. *Bioorganic and Medicinal Chemistry*, 25(17), 4637–4648. https://doi.org/10.1016/j.bmc.2017.06.046.

Kulkarni, A. A. (2014). Continuous flow nitration in miniaturized devices. *Beilstein Journal of Organic Chemistry*, 10, 405–424. https://doi.org/10.3762/bjoc.10.38.

Kumar, B., Sharma, P., Gupta, V. P., Khullar, M., Singh, S., Dogra, N., and Kumar, V. (2018). Synthesis and biological evaluation of pyrimidine bridged combretastatin derivatives as potential anticancer agents and mechanistic studies. *Bioorganic Chemistry*, 78, 130–140. https://doi.org/10.1016/j.bioorg.2018.02.027.

Li, N., Huang, J., Wang, Q., Gu, Y., and Wang, P. (2018). A reaction based one- and two-photon fluorescent probe for selective imaging H2O2 in living cells and tissues.

*Sensors and Actuators, B: Chemical*, 254, 411–416. https://doi.org/10.1016/j.snb.2017.07.133.

Lindstrom, U. M. (2007). *Organic Reactions in Water: Principles; Strategies and Applications*: Blackwell, Oxford, UK.

M., L. I. (2005). *Pyrimidine as Constituent of Natural Biologically Active Compounds. Chem. Biodivers.*, 2(1) 1-50.

Mahfoudh, M., Abderrahim, R., Leclerc, E., and Campagne, J. M. (2017). Recent Approaches to the Synthesis of Pyrimidine Derivatives. *European Journal of Organic Chemistry*, *2017*(20), 2856–2865. https://doi.org/10.1002/ejoc.201700008.

Manuscript, A. (2020). *Materials Chemistry B*. https://doi.org/10.1039/C9TB02528F.

Martinussen, J., and Hammer, K. (1995). Powerful methods to establish chromosomal markers in Lactococcus lactis: An analysis of pyrimidine salvage pathway mutants obtained by positive selections. *Microbiology*, 141(8), 1883–1890. https://doi.org/10.1099/13500872-141-8-1883.

Mastalir, M., Glatz, M., Pittenauer, E., Allmaier, G., and Kirchner, K. (2016). Sustainable Synthesis of Quinolines and Pyrimidines Catalyzed by Manganese PNP Pincer Complexes. *Journal of the American Chemical Society*, 138(48), 15543–15546. https://doi.org/10.1021/jacs.6b10433.

Motamedi, A., Sattari, E., Mirzaei, P., Armaghan, M., and Bazgir, A. (2014). An efficient and green synthesis of phthalide-fused pyrazole and pyrimidine derivatives. *Tetrahedron Letters*, 55(15), 2366–2368. https://doi.org/10.1016/j.tetlet.2014.02.101.

Naik, T. A., and Chikhalia, K. H. (2007). Studies on synthesis of pyrimidine derivatives and their pharmacological evaluation. *E-Journal of Chemistry*, 4(1), 60–66. https://doi.org/10.1155/2007/507590.

Nerkar, A. U. (2021). Use of Pyrimidine and Its Derivative in Pharmaceuticals: A Review. *Journal of Advanced Chemical Sciences*, 7(2), 729–732. https://doi.org/10.30799/jacs.239.21070203.

P. S. Baran, R. A. Shenvi, S. A. Nguyen, H. (2006). *No Title*. 70, 581-586.

Panneer Selvam, T., Richa James, C., Vijaysarathy Dniandev, P., and Karyn Valzita, S. (2012). A mini review of pyrimidine and fused pyrimidine marketed drugs. *Research in Pharmacy*, 2(4), 1–9.

Patil, S. B. (2023). Recent medicinal approaches of novel pyrimidine analogs: A review. *Heliyon*, 9(6), e16773. https://doi.org/10.1016/j.heliyon.2023.e16773.

Perupogu, N., Kumar, D. R., and Ramachandran, D. (2020). Anticancer activity of newly synthesized 1,2,4-Oxadiazole linked 4- (Oxazolo[5,4-d]pyrimidine derivatives. *Chemical Data Collections*, 27, 100363. https://doi.org/10.1016/j.cdc.2020.100363.

Qiu, X., Xin, C., Qin, W., Li, Z., Zhang, D., Zhang, G., Peng, B., Han, X., Yu, C., Li, L., and Huang, W. (2019). A novel pyrimidine based deep-red fluorogenic probe for detecting hydrogen peroxide in Parkinson's disease models. *Talanta*, 199(December 2018), 628–633. https://doi.org/10.1016/j.talanta.2019.03.017.

Safaei-Ghomi, J., and Ghasemzadeh, M. A. (2011). Ultrasound-assisted synthesis of dihydropyrimidine-2-thiones. *Journal of the Serbian Chemical Society*, 76(5), 679–684. https://doi.org/10.2298/JSC100212057S.

Schultz, D. C., Johnson, R. M., Ayyanathan, K., Miller, J., Whig, K., Kamalia, B., Dittmar, M., Weston, S., Hammond, H. L., Dillen, C., Ardanuy, J., Taylor, L., Lee, J. S., Li,

M., Lee, E., Shoffler, C., Petucci, C., Constant, S., Ferrer, M., … Cherry, S. (2022). Pyrimidine inhibitors synergize with nucleoside analogues to block SARS-CoV-2. *Nature*, 604(7904), 134–140. https://doi.org/10.1038/s41586-022-04482-x.

Sharma, V., Chitranshi, N., and Agarwal, A. K. (2014). Significance and Biological Importance of Pyrimidine in the Microbial World. *International Journal of Medicinal Chemistry*, 2014, 1–31. https://doi.org/10.1155/2014/202784.

Shi, X. Y., and Li, C. J. (2012). *Adv. Synth. Catal*., 354, 2933.

Shilpa, C., Dipak, S., Vimukta, S., and Arti, D. (2012). Comparative study of microwave and conventional synthesis and pharmacological activity of pyrimidines: A review. *International Journal of Pharmaceutical Sciences Review and Research*, 15(1), 15–22.

Shu, L., Chen, C., Huan, X., Huang, H., Wang, M., Zhang, J., Yan, Y., Liu, J., Zhang, T., and Zhang, D. (2020). Design, synthesis, and pharmacological evaluation of 4- or 6-phenyl-pyrimidine derivatives as novel and selective Janus kinase 3 inhibitors. *European Journal of Medicinal Chemistry*, 191(August 2019), 112148. https://doi.org/10.1016/j.ejmech.2020.112148.

Siddiqui, A. A., Rajesh, R., Mojahid-Ul-Islam, Alagarsamy, V., and De Clercq, E. (2007). Synthesis, antiviral, antituberculostic, and antibacterial activities of some novel, 4-(4-substituted phenyl)-6-(4-nitrophenyl)-2-(substituted imino)pyrimidines. *Archiv Der Pharmazie*, 340(2), 95–102. https://doi.org/10.1002/ardp.200600121.

Tan, Y. M., Li, D., Li, F. F., Fawad Ansari, M., Fang, B., and Zhou, C. H. (2022). Pyrimidine-conjugated fluoroquinolones as new potential broad-spectrum antibacterial agents. *Bioorganic and Medicinal Chemistry Letters*, 73. https://doi.org/10.1016/j.bmcl.2022.128885.

Terms, K. E. Y. (2019). *15.5: Aromatic Heterocycles - Pyridine and Pyrrole*. 4–6.

Tolba, M. S., Kamal El-Dean, A. M., Ahmed, M., Hassanien, R., Sayed, M., Zaki, R. M., Mohamed, S. K., Zawam, S. A., and Abdel-Raheem, S. A. A. (2022). Synthesis, reactions, and applications of pyrimidine derivatives. *Current Chemistry Letters*, 11(1), 121–138. https://doi.org/10.5267/J.CCL.2021.8.002.

Undheim, K. and T. B. (1996). Pyrimidines and their Benzo Derivatives. In: *Compre Hensive Heterocyclic Chemistry II*, Boulton, A.J. (Ed.)., Vol. 6, Elsevier Science Ltd., Oxford, UK., pp. 96-231.

Vadivel, M., Rajesh, J., Jeyamurugan, R., and Senthil, R. (2017). New pyrimidine based ligand capped gold and platinum nano particles: Synthesis, characterization, antimicrobial, antioxidant, DNA interaction and *in vitro* anticancer activities. *Journal of Photochemistry & Photobiology, B: Biology*. https://doi.org/10.1016/j.jphotobiol.2017.09.013.

Zahedifar, M., and Sheibani, H. (2015). Rapid three-component synthesis of pyrimidine and pyrimidinone derivatives in the presence of Bi(NO3)3·5H2O as a mild and highly efficient catalyst. *Research on Chemical Intermediates*, 41(1), 105–111. https://doi.org/10.1007/s11164-013-1172-6.

## Chapter 5

# The Use of Pyrimidine Derivatives as Adsorbents for the Removal and Extraction of Heavy Metal Ions

**Osman Çaylak**[1]
**and Burak Tüzün**[2,*]
[1]Department of Pharmacy Services, Vocational School of Health, Sivas Cumhuriyet University, Sivas, Turkey
[2]Plant and Animal Production Department, Technical Sciences Vocational School of Sivas, Sivas Cumhuriyet University, Sivas, Turkey

## Abstract

Pyrimidine and pyrimidine derivatives, which are included in heterocyclic compounds, are used in various applications in fields such as pharmaceutical medicine, biochemistry, analytical chemistry, polymer chemistry, coordination chemistry, and agricultural chemistry (drugs, organic compounds, polymers, sensors, inhibitors, etc., adsorbent etc.) This study presents the applications of pyrimidine derivatives in the adsorption of heavy metals, which are environmental pollutants. Studies have shown that pyrimidine derivatives are successfully applied for the enrichment and removal of toxic metals. Theoretical calculations of some molecules used in the study were calculated in the HF/6-31++G(d,p) basis set and the activities of the molecules were compared.

**Keywords:** pyrimidine derivatives, heavy metals, adsorption, removal, DFT

* Correspondence to: Assoc. Professor Dr. Burak Tüzün, Email: theburaktuzun@yahoo.com, Tel: +90 507 236 5991. http://orcid.org/ 0000-0002-0420-2043.

In: The Chemistry of Pyrimidine Derivatives
Editor: Barry Schneider
ISBN: 979-8-89113-563-5

# 1. Introduction

Almost all heavy metals, which are industrial trade substances that make our lives easier due to their chemical and physical properties, have toxic and carcinogenic effects [1]. Trace levels of some heavy metals are extremely important for the metabolic activities of living organisms. The release of heavy metals used in industrial activities into the environment through various means causes pollution in water and soil at a level that threatens living things. These pollutants, which are added to the food chain and habitat, enter the living body through various means and have a bio toxic effect by interacting with biomolecules such as proteins and enzymes. Thus, they inhibit biochemical reactions [2].

Metal pollution has become the riskiest environmental problem of our time. The removal of these pollutants, especially from wastewater, or their recovery, when necessary, is very important. There are numerous studies in the literature on this subject. In this section, we will touch upon some of the uses of pyrimidine derivatives for these purposes.

# 2. Adsorption of Heavy Metals by Pyrimidine Derivatives

Alireza Banaei et al., succeeded in removing silver and lead ions with high selectivity from a solution containing various metal ions with their newly synthesized bispyrimidine and bispyrazoline adsorbents [3]. The chemical structure of the bispyrimidine they synthesized is given in Figure 1.

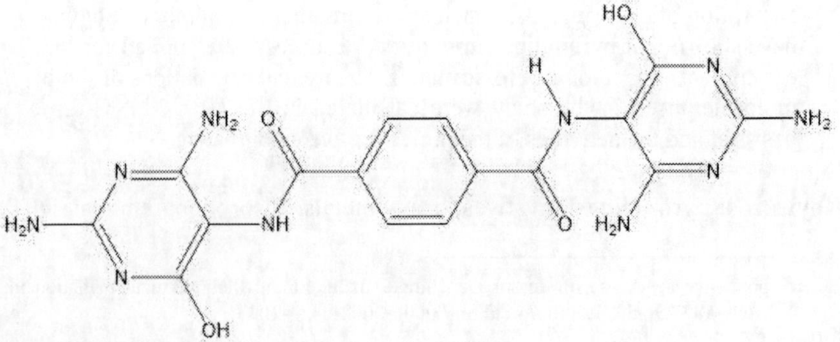
HO
H
N
N
NH2
NH2
O
N
N
H2N
NH
O
H2N
N
OH

**Figure 1.** Bispyrimidine.

Bulent Kirkan and Gul Asiye Aycik reported that Th(IV) ion from aqueous medium was enriched with good selectivity by adsorbent 5-[(E)-(5-sulfonyl-1,3,4-thiadiazol-2-yl) diazenil] pyrimidine-2,4,6 (1H, 3H, 5H)-trione Figure 2 [4].

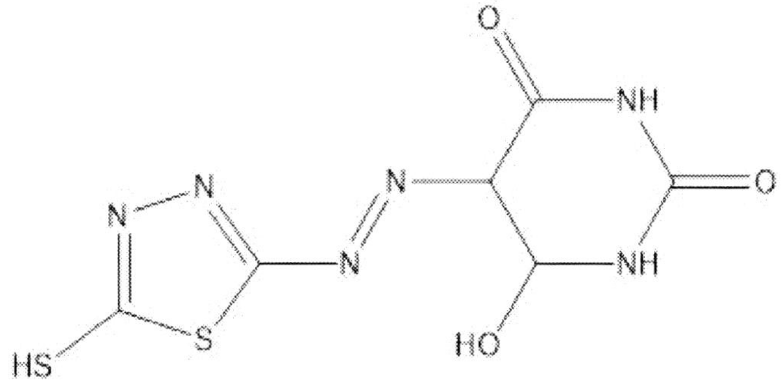
O
NH
N
N
O
N
N
NH
HS
S
HO

**Figure 2.** 5-[(E)-(5-sulfonyl-1,3,4-thiadiazol-2-yl) diazenil] pyrimidine-2,4,6 (1H, 3H, 5H)-trione.

Dingshuai Xue et al., modified polyurethane foam (PUF) with cytosine (Cyt) (Figure 3), a pyrimidine derivative, and used it as a column-filling material. Using this adsorbent, they developed a new solid phase extraction method for the separation and enrichment of trace levels of gold cations in various geological samples [5].

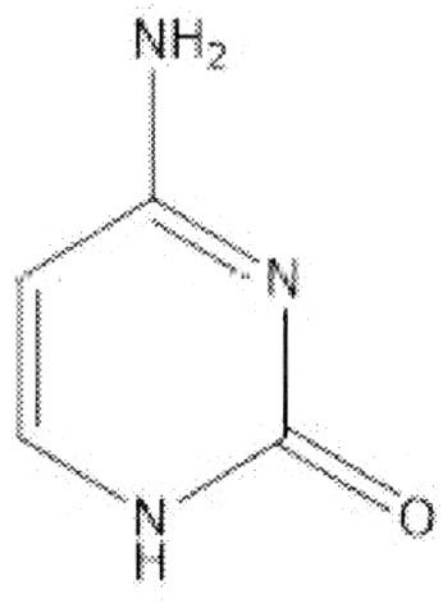
NH2
N
N
H
O

**Figure 3.** Cytosine.

Fatemeh Sabermahani et al., synthesized a new sorbent by coating silica gel with allyl 6-methyl-4-phenyl-2-thioxo-1,2,3,4-tetrahydropyrimidine-5-carboxylate pyrimidine derivative compound in Figure 4. They successfully applied this pyrimidine-derived adsorbent to the enrichment of trace levels of Pb(II) cations in various environmental samples [6].

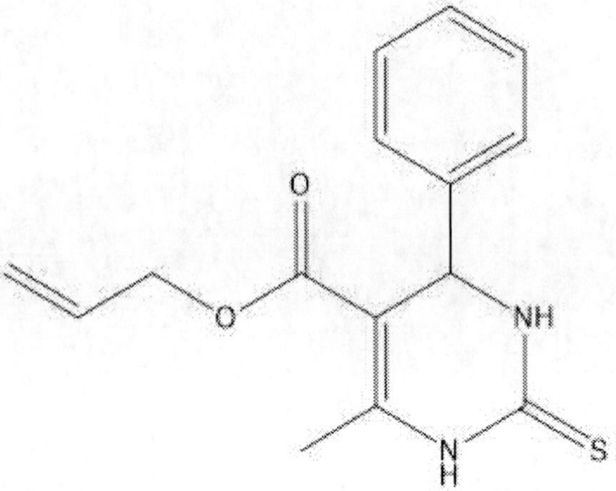
O
O
NH
N
H
S

**Figure 4.** Allyl 6-methyl-4-phenyl-2-thioxo-1,2,3,4-tetrahydropyrimidine-5-carboxylate.

Hamedreza Javadian et al., synthesized a methine-thiophene and pyrimidine-linked polyamide-derived polymer. The synthesized polymer and its calcium alginate-modified form were successfully used for the adsorption of Nd(III), Tb(III) and Dy(III) ions [7, 8].

Reduced graphene oxide-thymine composite was synthesized by Lingling Liu et al., by modifying the surface of reduced graphene oxide with thymine, a pyrimidine derivative in Figure 5. The resulting composite allowed the selective adsorption of the Hg(II) cation as well as various heavy metals. The affinity of thymine for Hg(II) provides a twofold advantage over reduced graphene oxide [9].

Maram H. Zahra et al., modified the surface of magnetic chitosan with 2-thioxodihydropyrimidine-4,6(1H,5H)-dione, a pyrimidine derivative. They reported that the synthesized pyrimidine derivative sorbent removed almost all of the Cr(VI) ions from highly polluted tannery wastewater with high affinity [10].

Maria D. Gutierrez-Valero et al., modified activated carbon by adsorption with pyrimidine-derived ligands. Activated carbon-pyrimidine adsorbents showed higher affinity for Cu(II) adsorption than activated carbon [11, 12].

The surface of silica gel was modified with 5-benzylidene-2-thiobarbituric acid in Figure 6, a pyrimidine derivative, by Mohamed E.

Mahmoud et al., it has been reported that the obtained silica gel-pyrimidine derivative adsorbent selectively adsorbed Cu(II), Hg(II), Cd(II) and Pb(II) heavy metal ions [13].

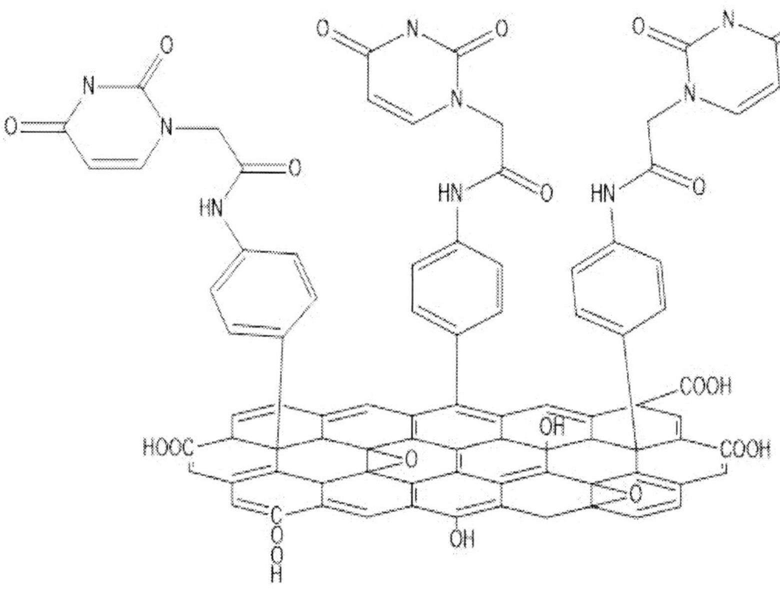

O
N
O
N
O
HN
O
N
O
N
HN
O
O
N
O
N
HN
O
COOH
OH
HOOC
O
COOH
O
C
O
O
H
OH


**Figure 5.** Reduced graphene oxide-thymine composites.

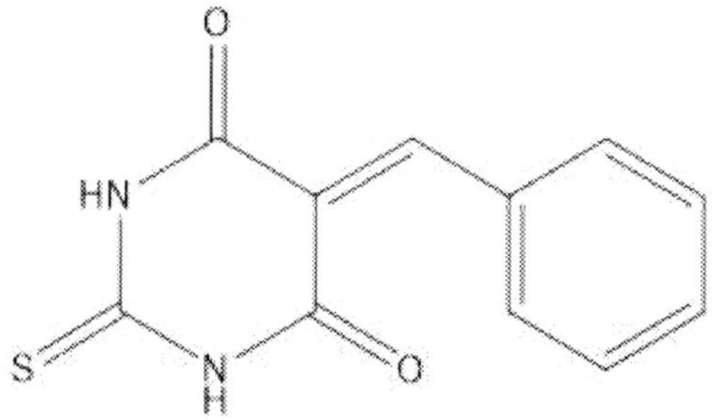

O
HN
S
N
H
O


**Figure 6.** 5-benzylidene-2-thiobarbituric acid.

Mohammed F. Hamza et al., obtained MC-PYS and MC-PYO sorbents by functionalizing magnetic chitosan microparticles with pyrimidine derivatives PYO and PYS in Figure 7. They successfully applied these sorbents to the recovery of silver ions in the wastewater of photographic films [14].

Mohammed F. Hamza et al., functionalized chitosan nanoparticles with 2-thioxodihydropyrimidine-4,6(1H,5H)-dione, a pyrimidine derivative. The obtained pyrimidine derivative sorbent was successfully applied to the removal of Cr(VI) cations from tannery wastes with high heavy metal load. They reported that this adsorbent may be suitable for water purification and chromium removal [15].

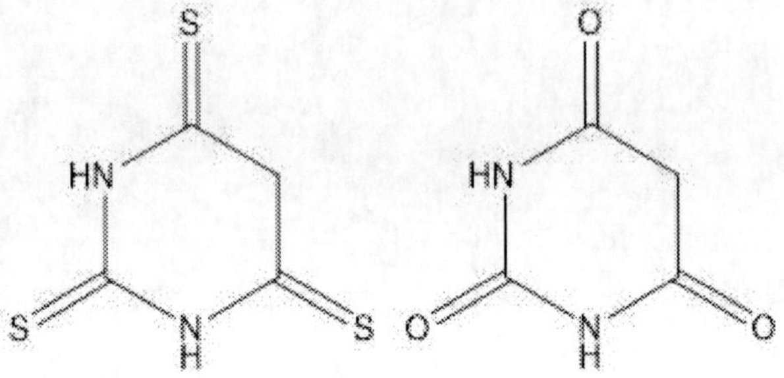
S
HN
S
N
H
S
O
HN
O
N
H
O

**Figure 7.** PYS and PYO molecules respectively.

Mohammed F. Hamza et al., obtained a new pyrimidine-derived adsorbent by reacting thiocarbazide and 2-thiobarbituric acid in Figure 8, with farmaldehyde. This adsorbent has been successfully applied in the removal of indium(III) in ore water with high heavy metal load [16].

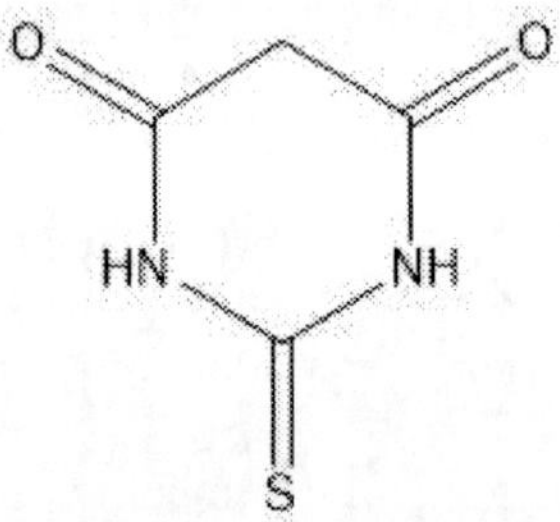
O
O
HN
NH
S

**Figure 8.** 2-Thioxodihydropyrimidine-4,6(1H,5H)-dione.

Phuong Lan Tran-Nguyen et al., synthesized a superparamagnetic thiamine/$Fe_3O_4$ adsorbent by functionalizing $Fe_3O_4$ with 3-((4-amino-2-methylpyrimidin-5-yl)methyl)-5-(2-hydroxyethyl)-4-methylthiazol-3-ium and a pyridine-derived compound in Figure 9. This adsorbent has an adsorption capacity for Cu(II) ions that is approximately six times higher than the $Fe_3O_4$ adsorbent. The study showed the positive effect of the pyrimidine ring on the adsorption capacity [17].

Sayed Zia Mohammadi et al., developed a solid phase extraction method based on the use of 5-(4-chloro-benzylidene) 1,3-dimethyl-pyrimidine-2,4,6 trione, a pyrimidine-derived compound, as a chelating agent. In the method, Co(II), Cu(II) and Mn(II) ions were enriched and successfully applied to real water samples [18].

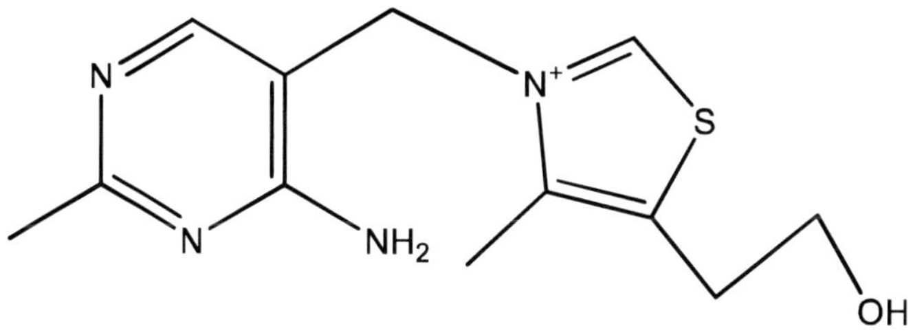
N
N
NH2
N+
S
OH

**Figure 9.** 3-((4-amino-2-methylpyrimidin-5-yl)methyl)-5-(2-hydroxyethyl)-4-methylthiazol-3-ium.

Seyed Jamilaldin Fatemi et al., developed a solid phase extraction method based on the extraction of thallium cations by modifying the surface of graphene oxide with 4-aminothieno[2,3-d] pyrimidine-2-thiol in Figure 10, a pyrimidine-derived compound. The method has been successfully applied to urine and various water samples [19].

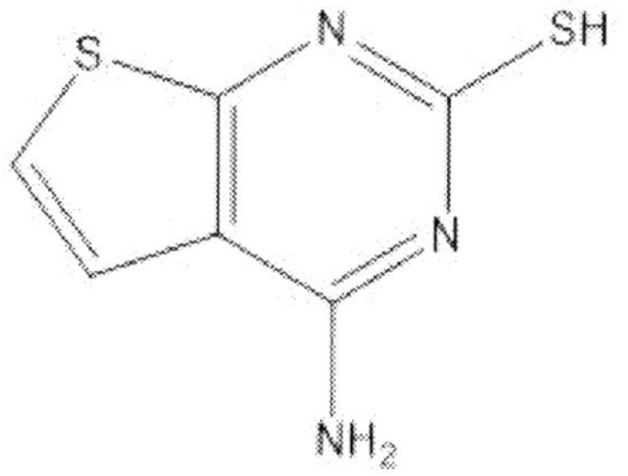
S
N
SH
N
NH2

**Figure 10.** 4-Aminothieno[2,3-d] pyrimidine-2-thiol.

M. Luz Godino-Salido et al., chose two different commercial activated carbon products obtained from different companies as starting materials. Two new hybrid materials were obtained by adsorbing pyrimidine-desferrioxamine-B conjugate compound ($H_4L$) in Figure 11 noncovalently on these activated carbons. The new adsorbents exhibited large adsorption capacity for Cu(II) ions. Moreover, these adsorbents allowed good recovery of Cu(II) ions [20].

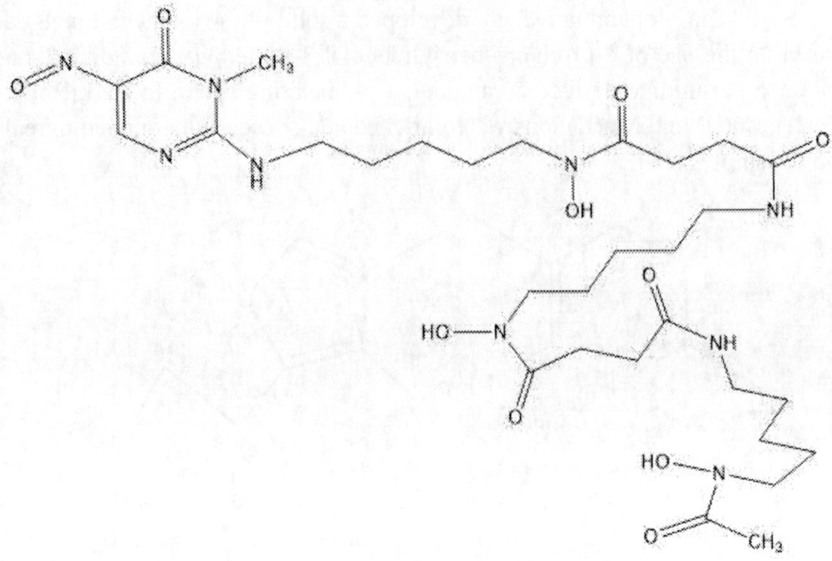

O
N
CH3
N
N
H
O
N
OH
O
NH
O
HO
N
NH
O
HO
N
O
CH3


**Figure 11.** Pyrimidine-desferrioxamine-B conjugate compound ($H_4L$).

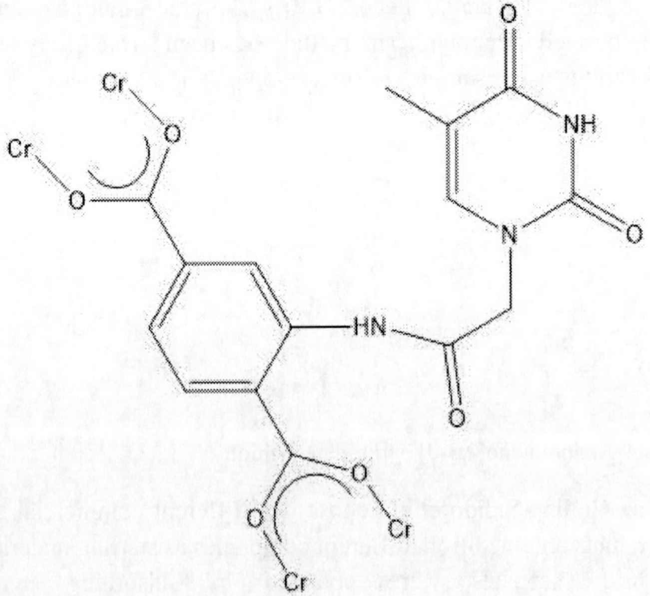

O
Cr
NH
Cr
O
O
N
O
HN
O
O
O
Cr
Cr


**Figure 12.** MIL-101-Timin.

Xubiao Luo et al., inspired by the interaction of thymine, a pyrimidine derivative, with Hg(II) and metal-organic frameworks, synthesized the thymine-functionalized MIL-101-Thymine in Figure 12, adsorbent. The adsorbent showed a high degree of selectivity towards Hg(II) ions, thanks to the thymine groups it contains. In addition, this adsorbent with its high adsorption capacity for mercury has been successfully applied to the removal of mercury in real water samples [21].

Yue Sun et al., functionalized polystyrene divinyl benzene beads with orotic acid in Figure 13, a pyrimidine-derived compound. The obtained new adsorbent was successfully applied to the removal of copper (II) ions from aqueous solution. The adsorbent exhibits a good adsorption capacity for copper ions and is reproducible with ammonia [22].

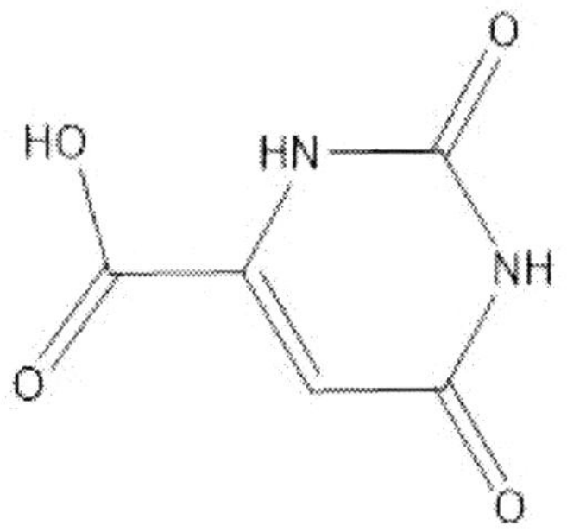
HO
HN
O
NH
O
O

**Figure 13.** Orotic acid.

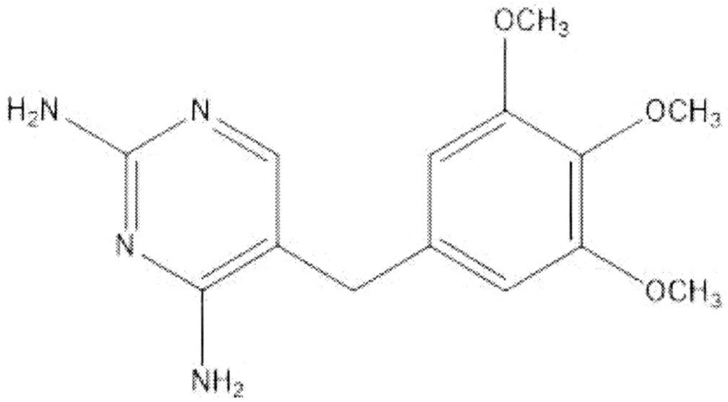
OCH3
H2N
N
OCH3
N
OCH3
NH2

**Figure 14.** 5-(3,4,5-trimethoxybenzyl)pyrimidine-2,4-diamine.

By Ali. H. Samir et al., the surface of magnetically reduced graphene oxide was modified with 5-(3,4,5-trimethoxybenzyl) pyrimidine-2,4-diamine in Figure 14, which is a pyrimidine-derived compound. The synthesized adsorbent was used to remove cobalt(II), nickel(II) and copper(II) ions from aqueous solution. The adsorbent has a good adsorption capacity for these heavy metal ions [23].

Thiazolylazopyrimidine-functionalized $TiO_2$ was synthesized by Zeinab Ghasemi and Asadollah Mohammadi using $TiO_2$ nanoparticles, 2,4,6-triaminopyrimidine in Figure 15, a pyrimidine-derived compound. The synthesized adsorbent has been successfully used in the sensitive and selective colorimetric determination of Cu(II) ions in various environmental water samples. [24].

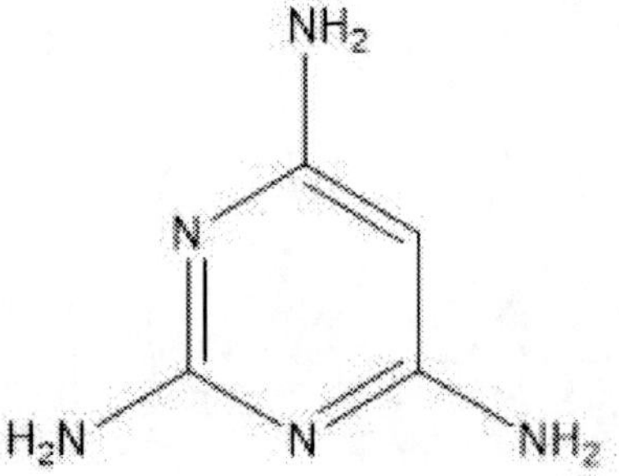
NH2
N
H2N
N
NH2

**Figure 15.** 2,4,6-Triaminopyrimidine.

As a result of theoretical calculations, parameters about many properties of molecules are calculated. Each parameter provides information about different properties of molecules. It is one of the fastest and easiest methods used to determine the activities and active sites of molecules through theoretical calculations [25]. These computational methods are used to identify the active sites of molecules, compare their activities, and examine activity changes against various biological materials. Although many programs are used in the calculations, the Gaussian package program is the most common among them. With this program, many quantum chemical parameters are calculated, and comparison is made using the numerical values of these parameters [26]. Indeed, it is possible to assess the activities of molecules through comparisons, and the most prominent parameters for such evaluations are the HOMO (Highest Occupied Molecular Orbital) and LUMO (Lowest Unoccupied Molecular Orbital). These molecular parameters are instrumental in making activity comparisons by examining their numerical values [27].

**Table 2.** The calculated quantum chemical parameters of molecules

| | $E_{HOMO}$ | $E_{LUMO}$ | I | A | ΔE | η | μ | χ | PA | ω | ε | dipol | Energy |
|---|---|---|---|---|---|---|---|---|---|---|---|---|---|
| **HF/6-31g LEVEL** | | | | | | | | | | | | | |
| **1** | -9.9722 | 1.0433 | 9.9722 | -1.0433 | 11.0155 | 5.5078 | 0.1816 | 4.4645 | -4.4645 | 1.8094 | 0.5527 | 2.7852 | -42909.1771 |
| **2** | -8.9717 | 0.9241 | 8.9717 | -0.9241 | 9.8958 | 4.9479 | 0.2021 | 4.0238 | -4.0238 | 1.6361 | 0.6112 | 3.7852 | -29329.9385 |
| **3** | -8.5632 | 0.8915 | 8.5632 | -0.8915 | 9.4547 | 4.7273 | 0.2115 | 3.8359 | -3.8359 | 1.5563 | 0.6426 | 6.0500 | -33559.4511 |
| **4** | -9.4019 | 0.4767 | 9.4019 | -0.4767 | 9.8786 | 4.9393 | 0.2025 | 4.4626 | -4.4626 | 2.0159 | 0.4961 | 4.3403 | -20535.2220 |
| **5** | -9.6767 | 0.0161 | 9.6767 | -0.0161 | 9.6928 | 4.8464 | 0.2063 | 4.8303 | -4.8303 | 2.4072 | 0.4154 | 0.8321 | -39598.4574 |
| **6** | -12.3464 | 0.9834 | 12.346 | -0.9834 | 13.3299 | 6.6649 | 0.1500 | 5.6815 | -5.6815 | 2.4216 | 0.4130 | 0.3778 | -13260.7464 |
| 7 | -9.9603 | 0.9364 | 9.9603 | -0.9364 | 10.8966 | 5.4483 | 0.1835 | 4.5120 | -4.5120 | 1.8683 | 0.5353 | 1.3298 | -22039.9640 |
| **8** | -6.7629 | 1.0109 | 6.7629 | -1.0109 | 7.7738 | 3.8869 | 0.2573 | 2.8760 | -2.8760 | 1.0640 | 0.9398 | 1.7070 | -31408.5933 |
| **9** | -8.3760 | 0.7864 | 8.3760 | -0.7864 | 9.1624 | 4.5812 | 0.2183 | 3.7948 | -3.7948 | 1.5717 | 0.6363 | 5.6168 | -32339.5338 |
| **10** | -10.4460 | 0.8120 | 10.446 | -0.8120 | 11.2580 | 5.6290 | 0.1777 | 4.8170 | -4.8170 | 2.0611 | 0.4852 | 4.2097 | -26742.7471 |
| **11** | -8.4076 | 1.0447 | 8.4076 | -1.0447 | 9.4522 | 4.7261 | 0.2116 | 3.6815 | -3.6815 | 1.4339 | 0.6974 | 1.8939 | -11640.3318 |

The numerical value of the HOMO parameter of the molecules signifies their ability to donate electrons, whereas the numerical value of the LUMO parameter indicates their capability to accept electrons [28]. It is a well-established fact that the molecule with the most positive numerical value for the HOMO parameter typically exhibits higher activity compared to other molecules. Conversely, molecules with the most negative numerical values for the LUMO parameter tend to demonstrate greater activity than their counterparts [29].

The concept of chemical hardness, introduced by Pearson in 1963 [30] during his exploration of Lewis acids and bases, can be defined as a measure of how resistant a chemical species is to the deformation or polarization of its electron cloud [31]. This notion paved the way for the formulation of principles like HSAB (Hard and Soft Acid-Base) [32] put forward by Drago and Kabler in 1972, and the PMH (Maximum Hardness Principle) [33] introduced by Pearson in 1993. These principles, rooted in the concept of chemical hardness, find applications in various theoretical and experimental studies.

Following the HSAB Principle [34], Lewis acids and bases are categorized as hard or soft. It posits that hard acids tend to preferentially interact with hard bases, while soft acids have an inclination for soft bases [35].

It's crucial to understand that 'soft' refers to polarizable chemical species, while 'hard' represents non-polarizable ones. It's widely recognized that polarizable species readily donate electrons to other molecules, while non-polarizable species face greater challenges in electron donation [36].

Chemical hardness of molecules is intricately linked to the energy gap (ΔE), as elucidated by Koopman's theorem. Hard molecules are distinguished by a significant HOMO-LUMO gap, while soft molecules feature a narrower HOMO-LUMO gap [37, 38]. Chemical hardness and softness are pivotal properties utilized in evaluating reactivity and stability. Softness (represented as 1/n), the reciprocal of chemical hardness, serves as an indicator of a molecule's polarizability [39, 40]. Soft molecules, known for their higher reactivity compared to hard counterparts, readily engage in electron donation to acceptors [40, 41].

While a multitude of quantum chemical parameters were calculated during the course of these computations, Figure 1 visualizes only a select few of them [42]. These visual representations play a crucial role in predicting the active sites within the molecules.

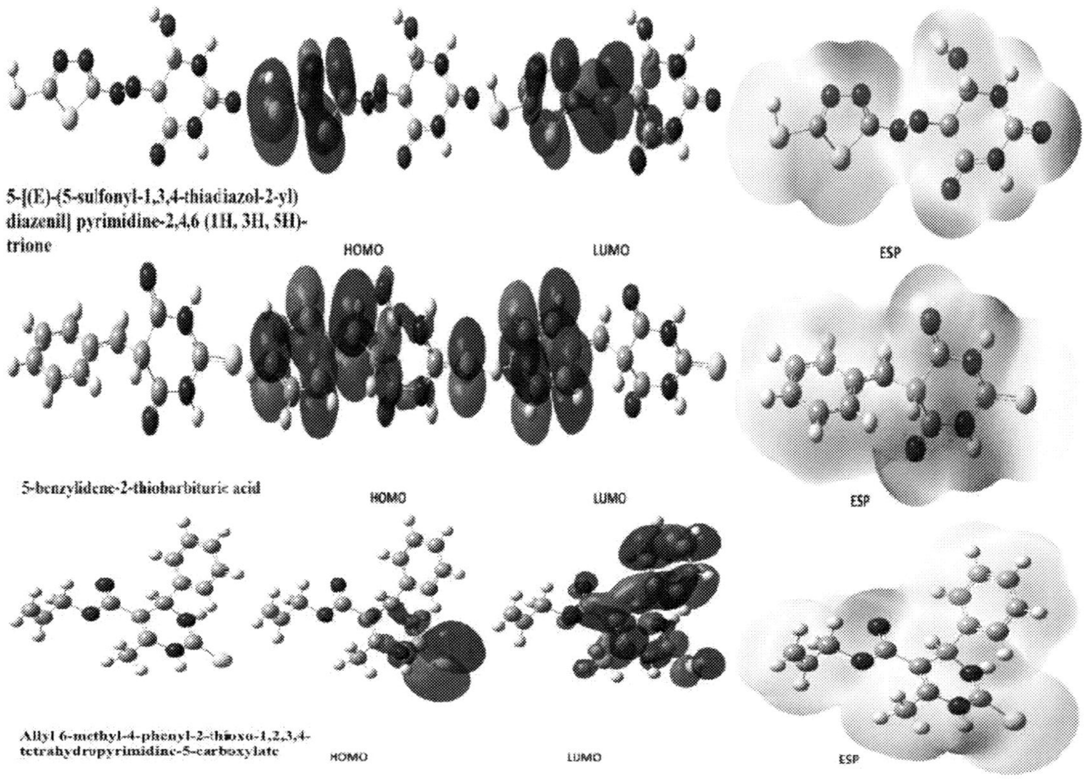
5-[(E)-(5-sulfonyl-1,3,4-thiadiazol-2-yl) diazenil] pyrimidine-2,4,6 (1H, 3H, 5H)-trione
HOMO
LUMO
ESP
5-benzylidene-2-thiobarbituric acid
HOMO
LUMO
ESP
Allyl 6-methyl-4-phenyl-2-thioxo-1,2,3,4-tetrahydropyrimidine-5-carboxylate
HOMO
LUMO
ESP

**Figure 16.** (Continued)

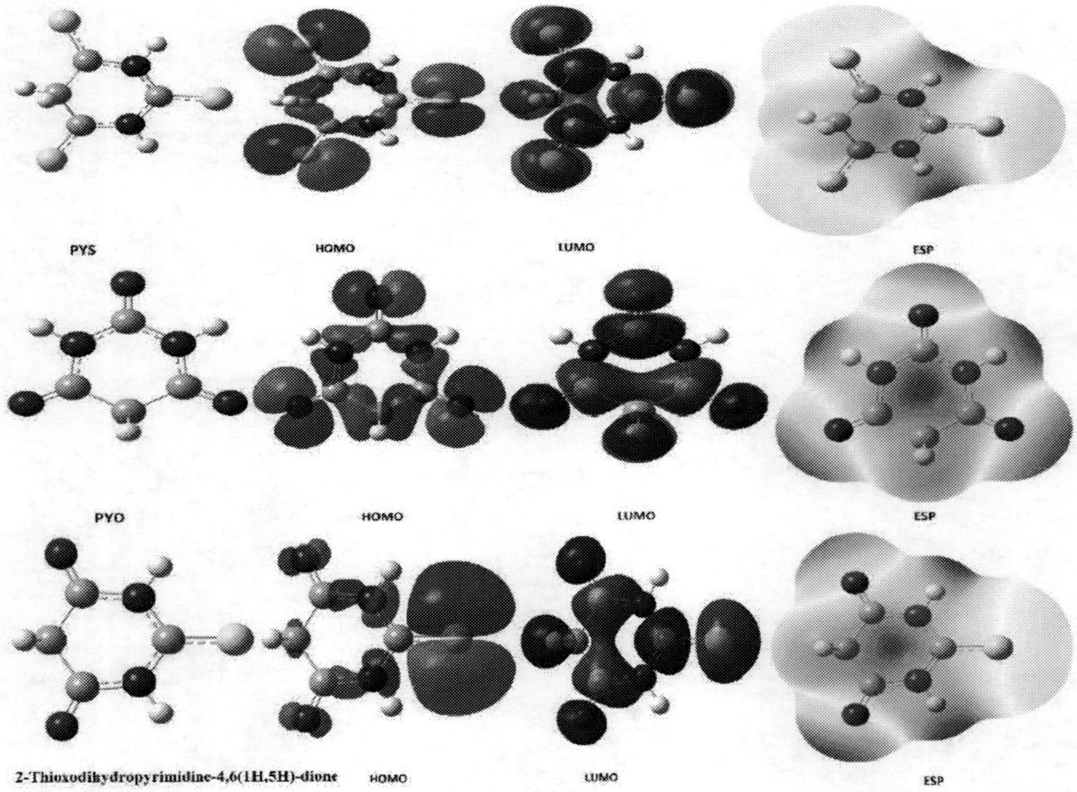
PYS
HOMO
LUMO
ESP
PYO
HOMO
LUMO
ESP
2-Thioxodihydropyrimidine-4,6(1H,5H)-dione
HOMO
LUMO
ESP

**Figure 16.** (Continued)

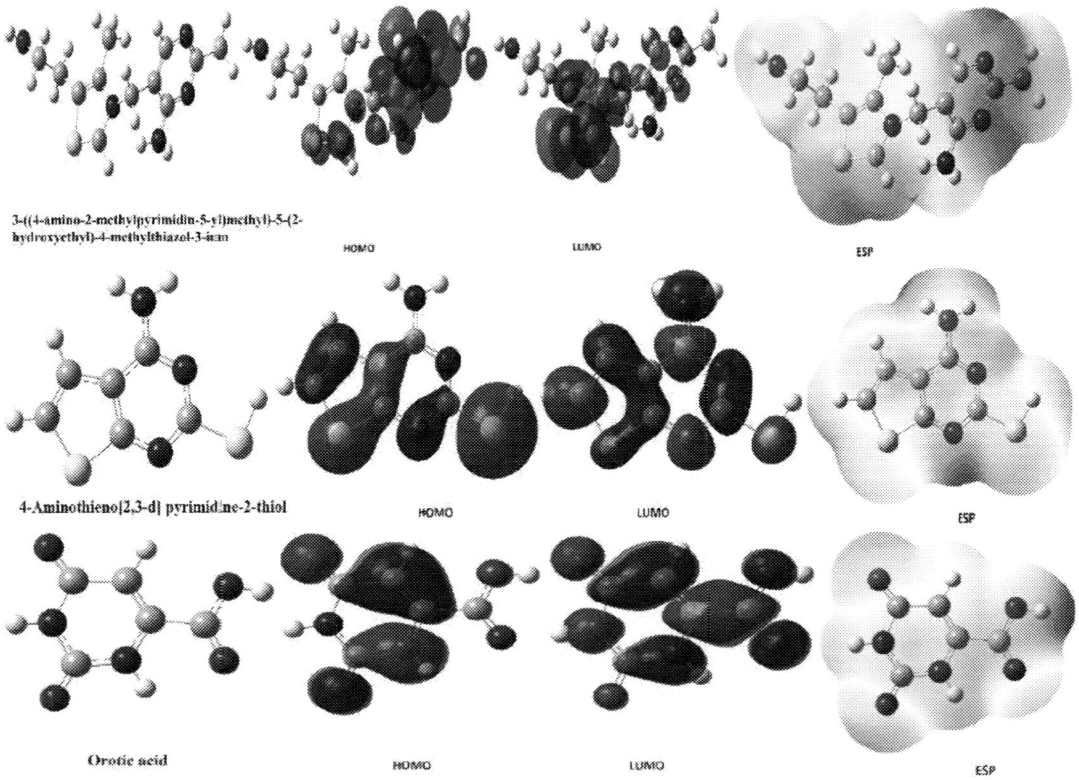
3-((4-amino-2-methylpyrimidin-5-yl)methyl)-5-(2-hydroxyethyl)-4-methylthiazol-3-ium
HOMO
LUMO
ESP
4-Aminothieno[2,3-d] pyrimidine-2-thiol
HOMO
LUMO
ESP
Orotic acid
HOMO
LUMO
ESP

**Figure 16.** (Continued)

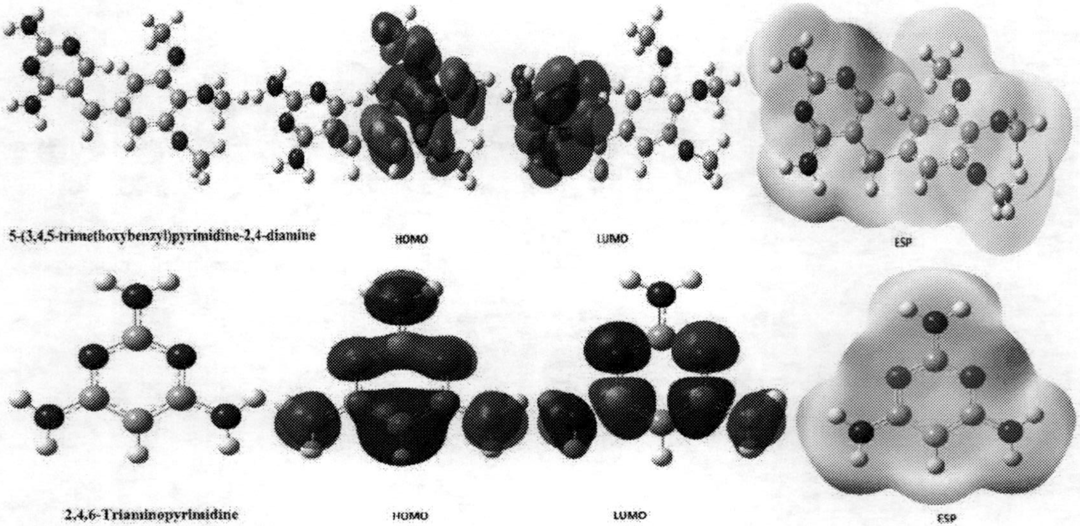
5-(3,4,5-trimethoxybenzyl)pyrimidine-2,4-diamine
HOMO
LUMO
ESP
2,4,6-Triaminopyrimidine
HOMO
LUMO
ESP

**Figure 16.** Representations of optimize structure, HOMO, LUMO, and ESP of some molecules.

In these calculations, the Electrostatic Potential (ESP) representations of the molecules offer valuable insights into the distribution of electrons within the molecule [43, 44]. These visualizations prominently feature regions with the highest electron distribution, which are marked in red, indicating areas where molecules readily engage in electron donation interactions [45]. In contrast, areas with the lowest electron distribution are represented in blue, signifying regions where molecules are more inclined to accept electrons [46, 47].

## Conclusion

The release of heavy metals, which are very dangerous for living things, into the environment and their concentrations increasing day by day have become a serious problem. It shows that pyrimidine-derived adsorbents synthesized for the removal/enrichment of heavy metal contaminants have been successfully applied to cope with pollution.

It is possible to compare the activities of pyrimidine derivatives with theoretical calculations. The numerical values of quantum chemical parameters obtained as a result of the calculations provide a lot of information about pyrimidine derivative molecules. According to the numerical value of the HOMO parameters of the molecules, the 4-Aminothieno[2,3-d] pyrimidine-2-thiol molecule in this study has the highest activity according to the value of -6.7629. On the other hand, the PYO molecule was calculated to be the most active molecule with the numerical value of the LUMO parameter being 0.0161. The calculations provided important information to compare the activities of the molecules.

## Acknowledgments

The numerical calculations reported in this paper were fully/partially performed at TUBITAK ULAKBIM, High Performance and Grid Computing Center (TRUBA resources).

## References


[1] Kim, H. S., Kim, Y. J., & Seo, Y. R. (2015). An overview of carcinogenic heavy metal: molecular toxicity mechanism and prevention. *Journal of cancer prevention*, 20(4), 232.

[2] Mishra, S., Bharagava, R. N., More, N., Yadav, A., Zainith, S., Mani, S., & Chowdhary, P. (2019). Heavy metal contamination: an alarming threat to environment and human health. *Environmental biotechnology: For sustainable future*, 103-125.

[3] Banaei, A., Salmanpour, H., & Karimi, S. (2017). Ultrasound irradiation promoted synthesis of bispyrimidine and bispyrazolines as selective adsorbents for the selective removal of Ag (I) and Pb (II) from aqueous solutions. *Monatshefte für Chemie-Chemical Monthly*, *148*, 683-690.

[4] Kırkan, B., & Aycik, G. A. (2016). Solid phase extraction using silica gel modified with azo-dyes derivative for preconcentration and separation of Th (IV) ions from aqueous solutions. *Journal of Radioanalytical and Nuclear Chemistry*, *308*, 81-91.

[5] Xue, D., Wang, H., Liu, Y., Shen, P., & Sun, J. (2016). Cytosine-functionalized polyurethane foam and its use as a sorbent for the determination of gold in geological samples. *Analytical Methods*, *8*(1), 29-39.

[6] Sabermahani, F., Hassani, Z., & Faramarzpoor, M. (2013). The use of silica gel modified with allyl 6-methyl-4-phenyl-2-thioxo-1, 2, 3, 4-tetrahydropyrimidine-5-carboxylate for selective separation and preconcentration of lead in environmental samples. *Environmental monitoring and assessment*, *185*, 4969-4976.

[7] Javadian, H., Taghavi, M., Ruiz, M., Tyagi, I., Farsadrooh, M., & Sastre, A. M. (2022). Adsorption of neodymium, terbium and dysprosium using a synthetic polymer-based magnetic adsorbent. *Journal of Rare Earths*.

[8] Javadian, H., Ruiz, M., Taghvai, M., & Sastre, A. M. (2020). Novel magnetic nanocomposite of calcium alginate carrying poly (pyrimidine-thiophene-amide) as a novel green synthesized polyamide for adsorption study of neodymium, terbium, and dysprosium rare-earth ions. *Colloids and Surfaces A: Physicochemical and Engineering Aspects*, *603*, 125252.

[9] Liu, L., Ding, L., Wu, X., Deng, F., Kang, R., & Luo, X. (2016). Enhancing the Hg (II) removal efficiency from real wastewater by novel thymine-grafted reduced graphene oxide complexes. *Industrial & Engineering Chemistry Research*, *55*(24), 6845-6853.

[10] Zahra, M. H., Hamza, M. F., El-Habibi, G., Abdel-Rahman, A. A. H., Mira, H. I., Wei, Y., Alotaibi, S. H., Amer H. H., Goda, A. E.-S., & Hamad, N. A. (2022). Synthesis of a novel adsorbent based on chitosan magnetite nanoparticles for the high sorption of Cr (VI) ions: A study of photocatalysis and recovery on tannery effluents. *Catalysts*, *12*(7), 678.

[11] Gutiérrez-Valero, M. D., Godino-Salido, M. L., Arranz-Mascarós, P., López-Garzón, R., Cuesta, R., & García-Martín, J. (2007). Adsorption of designed pyrimidine derivative ligands on an activated carbon for the removal of Cu (II) ions from aqueous solution. *Langmuir*, *23*(11), 5995-6003.

[12] Gutiérrez-Valero, M. D., Arranz-Mascarós, P., Godino-Salido, M. L., López-León, M. D., López-Garzón, R., & Cuesta, R. (2008). Adsorption of a designed L-glutamic acid-pyrimidine derivative ligand on an activated carbon for the removal of Cu (II) from aqueous solution. *Microporous and mesoporous materials*, *116*(1-3), 445-451.

[13] Mahmoud, M. E., Haggag, S. S., & Hegazi, A. H. (2006). Synthesis, characterization, and sorption properties of silica gel-immobilized pyrimidine derivative. *Journal of colloid and interface Science*, *300*(1), 94-99.

[14] Hamza, M. F., Adel, A. H., Hawata, M. A., El Araby, R., Guibal, E., Fouda, A., Wei Y., & Hamad, N. A. (2022). Functionalization of magnetic chitosan microparticles–Comparison of trione and trithione grafting for enhanced silver sorption and application to metal recovery from waste X-ray photographic films. *Journal of Environmental Chemical Engineering*, *10*(3), 107939.

[15] Hamza, M. F., Wei, Y., Althumayri, K., Fouda, A., & Hamad, N. A. (2022). Synthesis and Characterization of Functionalized Chitosan Nanoparticles with Pyrimidine Derivative for Enhancing Ion Sorption and Application for Removal of Contaminants. *Materials*, *15*(13), 4676.

[16] Hamza, M. F., Abd Allh, M., Guibal, E., Abdel-Rahman, A. A., & El Araby, R. (2023). Synthesis of a new pyrimidine-based sorbent for indium (III) removal from aqueous solutions–Application to ore leachate. *Separation and Purification Technology*, *314*, 123514.

[17] Tran-Nguyen, P. L., Angkawijaya, A. E., Ha, Q. N., Tran-Chuong, Y. N., Go, A. W., Bundjaja, V., Gunarto, C., Santoso, S. P., & Ju, Y. H. (2022). Facile synthesis of superparamagnetic thiamine/Fe3O4 with enhanced adsorptivity toward divalent copper ions. *Chemosphere*, *291*, 132759.

[18] Mohammadi, S. Z., Hosseiny, E. S., & Moslemi, S. (2014). Preconcentration trace amounts of cobalt, copper and manganese ions using in-situ surfactant-based solid phase extraction prior to its FAAS determination.

[19] Fatemi, S. J., Akhgar, M. R., & Abbasabadi, M. K. (2021). Thallium extraction in urine and water samples by nanomagnetic 4-Aminothieno [2, 3-d] pyrimidine-2-thiol functionalized on graphene oxide. *Analytical Methods in Environmental Chemistry Journal*, *4*(03), 68-79.

[20] Godino-Salido, M. L., Santiago-Medina, A., López-Garzón, R., Gutiérrez-Valero, M. D., Arranz-Mascarós, P., de la Torre, M. D. L., Domingo-García, M., & López-Garzón, F. J. (2016). Preparation and characterization of trihydroxamic acid functionalized carbon materials for the removal of Cu (II) ions from aqueous solution. *Applied Surface Science*, *387*, 128-138.

[21] Luo, X., Shen, T., Ding, L., Zhong, W., Luo, J., & Luo, S. (2016). Novel thymine-functionalized MIL-101 prepared by post-synthesis and enhanced removal of Hg2+ from water. *Journal of hazardous materials*, *306*, 313-322.

[22] Sun, Y., Li, Z. C., & Xu, Y. (2013). Preparation and application of a novel orotic acid chelating resin for removal of Cu (II) in aqueous solutions. *Chinese Chemical Letters*, *24*(8), 747-750.

[23] Mohammed, Y. A., Al-Somaidaie, G. H., & Samir, A. H. (2020). Magnetic Reduced Graphene Oxide-5-(3, 4, 5 trimethoxybenzyl) pyrimidine-2, 4-yl (MRGO-Ar) for

the removal of Co (II), Ni (II) and Cu (II) ions from aqueous solution: Synthesis and Adsorption. *Journal of Education and Scientific Studies*, *1*(15).

[24] Ghasemi, Z., & Mohammadi, A. (2020). Sensitive and selective colorimetric detection of Cu (II) in water samples by thiazolylazopyrimidine-functionalized TiO2 nanoparticles. *Spectrochimica Acta Part A: Molecular and Biomolecular Spectroscopy*, *239*, 118554.

[25] Al Ati, G., Chkirate, K., El-Guourrami, O., Chakchak, H., Tüzün, B., Mague, J. T., Benzeid, H., Achour R., & Essassi, E. M. (2024). Schiff base compounds constructed from pyrazole–acetamide: Synthesis, spectroscopic characterization, crystal structure, DFT, molecular docking and antioxidant activity. *Journal of Molecular Structure*, 1295, 136637.

[26] Lakhrissi, Y., Rbaa, M., Tuzun, B., Hichar, A., Ounine, K., Almalki, F., Anouar, El H., Hadda, T. B., Zarrouk, A., & Lakhrissi, B. (2022). Synthesis, structural confirmation, antibacterial properties and bio-informatics computational analyses of new pyrrole based on 8-hydroxyquinoline. *Journal of Molecular Structure*, 1259, 132683.

[27] Lakhrissi, Y., Rbaa, M., Tuzun, B., Hichar, A., Ounine, K., Almalki, F., Anouar, El H., Hadda, T. B., Zarrouk, A., & Lakhrissi, B. (2022). Synthesis, structural confirmation, antibacterial properties and bio-informatics computational analyses of new pyrrole based on 8-hydroxyquinoline. *Journal of Molecular Structure*, 1259, 132683.

[28] Güzel, E., Günsel, A., Tüzün, B., Atmaca, G. Y., Bilgiçli, A. T., Erdoğmuş, A., & Yarasir, M. N. (2019). Synthesis of tetra-substituted metallophthalocyanines: Spectral, structural, computational studies and investigation of their photophysical and photochemical properties. *Polyhedron*, 158, 316-324.

[29] Günsel, A., Kobyaoğlu, A., Bilgicli, A. T., Tüzün, B., Tosun, B., Arabaci, G., & Yarasir, M. N. (2020). Novel biologically active metallophthalocyanines as promising antioxidant-antibacterial agents: Synthesis, characterization and computational properties. *Journal of Molecular Structure*, 1200, 127127.

[30] Pearson, R. G. (1963). Hard and soft acids and bases. *Journal of the American Chemical society*, 85(22), 3533-3539.

[31] Günsel, A., Kırbaç, E., Tüzün, B., Erdoğmuş, A., Bilgiçli, A. T., & Yarasir, M. N. (2019). Selective chemosensor phthalocyanines for Pd2+ ions; synthesis, characterization, quantum chemical calculation, photochemical and photophysical properties. *Journal of Molecular Structure*, 1180, 127-138.

[32] Drago, R. S., & Kabler, R. A. (1972). Quantitative evaluation of the HSAB [hard-soft acid-base] concept. *Inorganic Chemistry*, 11(12), 3144-3145.

[33] Toro-Labbé, A. (1999). Characterization of chemical reactions from the profiles of energy, chemical potential, and hardness. *The Journal of Physical Chemistry A*, 103(22), 4398-4403.

[34] Chattaraj, P. K., Lee, H., & Parr, R. G. (1991). HSAB principle. *Journal of the American Chemical Society*, 113(5), 1855-1856.

[35] Tüzün, B., & Sayin, K. (2019). Investigations over optical properties of boron complexes of benzothiazolines. *Spectrochimica Acta Part A: Molecular and Biomolecular Spectroscopy*, 208, 48-56.

[36] Kaya, S., Tüzün, B., Kaya, C., & Obot, I. B. (2016). Determination of corrosion inhibition effects of amino acids: quantum chemical and molecular dynamic simulation study. *Journal of the Taiwan Institute of Chemical Engineers*, 58, 528-535.

[37] Tüzün, B. (2020). Investigation of pyrazoly derivatives schiff base ligands and their metal complexes used as anti-cancer drug. *Spectrochimica Acta Part A: Molecular and Biomolecular Spectroscopy*, 227, 117663.

[38] Karrouchi, K., Fettach, S., Tüzün, B., Radi, S., Alharthi, A. I., Ghabbour, H. A., Anouar, E. H., Mabkhot, Y. N., Faouzi, M. E. A., Ansar, M., & Garcia, Y. (2021). Synthesis, crystal structure, DFT, α-glucosidase and α-amylase inhibition and molecular docking studies of (E)-N'-(4-chlorobenzylidene)-5-phenyl-1H-pyrazole-3-carbohydrazide. *Journal of Molecular Structure*, 1245, 131067.

[39] Kanzouai, Y., Chalkha, M., Hadni, H., Laghmari, M., Bouzammit, R., Nakkabi, A., Benali, T., Tüzün, B., Akhazzane, M., Yazidi, M. E., & Al Houari, G. (2023). Design, synthesis, in-vitro and in-silico studies of chromone-isoxazoline conjugates as anti-bacterial agents. *Journal of Molecular Structure*, 1293, 136205.

[40] Aksu, A., Çetinkaya, S., Yenidünya, A. F., Çetinus, Ş. A., Gezegen, H., & Tüzün, B. (2023). Immobilization of pectinase on chitosan-alginate-clay composite beads: Experimental, DFT and molecular docking studies. *Journal of Molecular Liquids*, 390, 122947.

[41] Bensalah, J., Ouaddari, H., Erdoğan, Ş., Tüzün, B., Gaafar, A. R. Z., Nafidi, H. A., Bourhia, M., & Habsaoui, A. (2023). Cationic resin polymer A® IRC-50 as an effective adsorbent for the removal of Cr (III), Cu (II), and Ag (I) from aqueous solutions: A kinetic, mathematical, thermodynamic and modeling study. *Inorganic Chemistry Communications*, 157, 111272.

[42] Erdoğan, M., Yeşildağ, A., Yıldız, B., Tüzün, B., & Özden, Ö. (2023). Synthesis and characterization of some benzidine-based azomethine derivatives with molecular docking studies and anticancer activities. *Chemical Papers*, 77(11), 6829-6847.

[43] Tuzun, B., Taş, N. A., Taslimi, P., & Karadağ, A. (2023). Synthesis, enzyme inhibition, and in silico studies of Amino Acid Schiff Bases. *Iranian Journal of Chemistry and Chemical Engineering*.

[44] Tapera, M., Kekeçmuhammed, H., Sarıpınar, E., Doğan, M., Tüzün, B., Koçyiğit, Ü. M., & Çetin, F. N. (2023). Molecular Hybrids Integrated with Imidazole and Hydrazone structural motifs: Design, Synthesis, Biological Evaluation, and Molecular Docking Studies. *Journal of Molecular Liquids*, 123242.

[45] Manap, S., Medetalibeyoğlu, H., Kılıç, A., Karataş, O. F., Tüzün, B., Alkan, M., Ortaakarsu, A. B., Atalay, A., Beytur, M., & Yüksek, H. (2023). Synthesis, molecular modeling investigation, molecular dynamic and ADME prediction of some novel Mannich bases derived from 1, 2, 4-triazole, and assessment of their anticancer activity. *Journal of Biomolecular Structure and Dynamics*, 1-15.

[46] Arukalam, I. O., Uzochukwu, I. N., Izionworu, V. O., Tüzün, B., & Dagdag, O. (2023). Corrosion protection of Q235 steel in Pseudomonas aeruginosa-laden seawater environment using high barrier PDMS nanocomposite coating. *Safety in Extreme Environments*, 1-11.

[47] Aksu, A., Çelik, M. S., Polat, Z. A., Yenidünya, A. F., & Tuzun, B. (2023). Experimental and theoretical evidence on the amoebicidal activity of synthesized tRNA-palmitic acid esters. *Iranian Journal of Chemistry and Chemical Engineering.*

Chapter 6

# Pyrimidine Derivatives: A Crucial Scaffold in Discovery of Anti-Cancer Agents

**Aadya Passi**
**Bhumi Baweja**
**Ronit Chakraborty**
**Abhishek Wahi**
**and Priti Jain***
Department of Pharmaceutical Chemistry, School of Pharmaceutical Sciences, Delhi Pharmaceutical Sciences and Research University (DPSRU), New Delhi, Delhi, India

## Abstract

Cancer poses a significant global health challenge due to its complex nature. Hence, medicinal chemists have made significant efforts in order to develop effective and safe anticancer drugs. With the advent of novel anticancer drugs having diverse heterocyclic rings. One such heterocyclic ring is pyrimidine. Pyrimidine derivatives' wide spectrum of biological and pharmacological properties has drawn a lot of attention in the field of medicinal chemistry. It has been found that pyrimidines, as well as pyrimidine-fused heterocyclic compounds, exhibit anticancer effects via a range of targets and pathways. Herein, this chapter explores the intricate domain of cancer and underscores the important role of pyrimidine derivatives in enhancing our understanding and treatment strategies. The chapter provides readers an assessment of the molecular targets linked to cancer, aiming to clarify their role in disease

* Corresponding Author's Email: pritijain@dpsru.edu.in.

In: The Chemistry of Pyrimidine Derivatives
Editor: Barry Schneider
ISBN: 979-8-89113-563-5

advancement and the possibility for treatment. It presents an extensive overview of the essential attributes and composition of pyrimidine, emphasizing its anti-cancer qualities while explaining the specific sites within cancer cells targeted by pyrimidine derivatives with respect to their pharmacology and medicinal chemistry. Significant commercial products containing pyrimidine derivatives are analyzed in detail, with a focus on their effectiveness and the leading companies behind these innovative formulations. All in all, it summarizes the notable discoveries, underscoring the vital role of pyrimidine derivatives in discovery of anticancer agents.

**Keywords:** cancer, pyrimidine, heterocyclic compounds, medicinal chemistry

## Introduction

Cancer encompasses a vast array of over 277 types of diseases. In 2018, there was a significant increase in global cancer diagnoses. Lung and prostate cancers were more common in men, while breast cancer was prevalent among women. Colorectal cancer ranked third globally and second most common in women. The age-standardized rates showed that breast cancer had the highest frequency, followed by prostate and lung cancers. The overall risk of developing cancer for certain age groups was notable, with specific risks related to gender such as lung and breast cancers (Mattiuzzi and Lippi 2019). Cancer arises from the transformation of normal cells into tumour cells, typically progressing through stages starting with pre-cancerous lesions and leading to the development of malignant tumors. This process is influenced by both genetic factors and external agents, which can be categorized as physical carcinogens (such as ultraviolet radiation), chemical carcinogens (including components of tobacco smoke, asbestos, alcohol, aflatoxin, and arsenic), and biological carcinogens (such as viral, bacterial or parasitic infections) (Hanahan and Weinberg 2011; Soto and Sonnenschein 2010).

Scientists have identified various stages of cancer, indicating that multiple gene mutations play a role in its development. These genetic mutations result in abnormal cell growth and proliferation. Inherited genetic disorders also contribute significantly to increased cellular growth. Advances in bioinformatics and molecular techniques have provided additional information that can aid in early diagnosis and effective treatment strategies (Malone et al., 2020). The impact of medications on cancer patients can be predicted and managed to mitigate certain side effects. Molecular genetic studies conducted in recent years have successfully revealed the mechanisms involved in carcinogenesis, leading to an enhanced understanding of how genetic disorders contribute to the formation of cancer cells (Liu et al., 2021). In the past few decades, there have been significant advancements in anticancer drug discovery. These include the identification of new targets, targeted therapies, and immunotherapy. High-throughput screening, molecular biology, and structure-based drug design have played a crucial role in expediting candidate identification. Alongside traditional agents, innovative approaches such as tyrosine kinase inhibitors are being explored. Despite these developments, challenges related to drug resistance and toxicity persist (Zhong et al., 2021). The integration of genomics, proteomics, and AI holds promise for personalized medicine in this ever-evolving field of anticancer drug discovery. Interdisciplinary collaboration and technological innovations continue to be key factors driving progress.

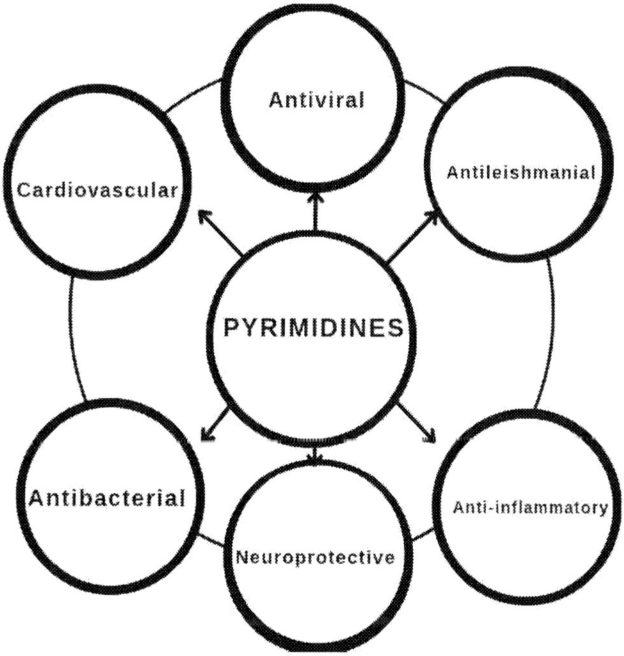
Antiviral
Cardiovascular
Antileishmanial
PYRIMIDINES
Antibacterial
Neuroprotective
Anti-inflammatory

**Figure 1.** Various biological activities of pyrimidine containing compounds.

**Table 1.** The biological targets, chemical structures, and applications of several FDA approved pyrimidine-containing medicinal candidates (https://go.drugbank.com/)

| S.No. | Name of drug candidate | Structure | Pharmacological class | Mechanism of inhibition | Application |
|---|---|---|---|---|---|
| 1. | Methotrexate | 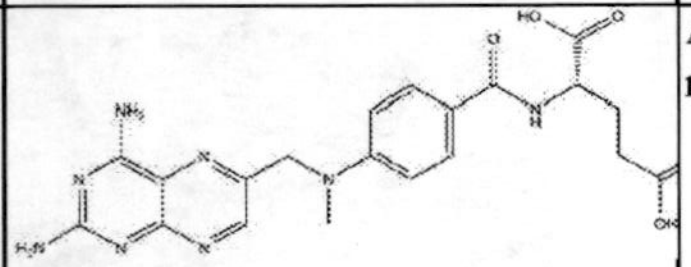 | Anti-folate metabolite | In cancer treatment, it acts as an antifolate antimetabolite by entering the cell through human reduced folate carriers and forms methotrexate-polyglutamate. Both methotrexate and methotrexate-polyglutamate inhibit the enzyme dihydrofolate reductase (DHFRase), which converts dihydrofolate into tetrahydrofolate - the active form of folic acid. Tetrahydrofolate is essential for synthesizing nucleotides for both DNA and RNA. | Acute lymphoblastic leukaemia, acute myeloid leukaemia, osteosarcomas, breast cancer |
| 2. | 5-fluorouracil | 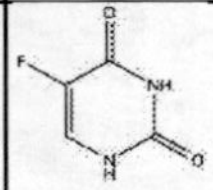 | Antimetabolite | The primary mechanism of action for fluorouracil involves the formation of a covalently bound complex between the deoxyribonucleotide of the drug and the folate cofactor, N5–10-methylenetetrahydrofolate, with thymidylate synthase. This results in inhibiting thymidylate formation from uracil, thereby hindering DNA and RNA synthesis leading to cell death. Additionally, fluorouracil can replace uridine triphosphate in RNA production, disrupting RNA processing and protein synthesis. | Colorectal Cancer, Basal Cell Carcinoma |

| S.No. | Name of drug candidate | Structure | Pharmacological class | Mechanism of inhibition | Application |
|---|---|---|---|---|---|
| 3. | Gemcitabine | 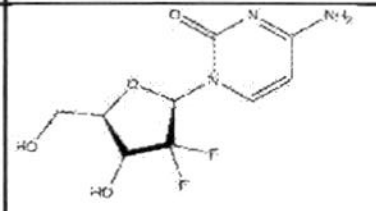 | Antimetabolite | Gemcitabine, a powerful analogue of deoxycytidine, is phosphorylated in cancer cells to form active compounds known as gemcitabine diphosphate (dFdCDP) and gemcitabine triphosphate (dFdCTP). These nucleosides competitively inhibit DNA chain elongation by competing with deoxycytidine triphosphate (dCTP), leading to termination of the DNA chain. This inclusion ultimately causes DNA fragmentation and cell death through apoptosis in malignant cells | Ovarian Cancer, Breast Cancer and Non-small cell Lung cancer |
| 4. | Trifluridine | 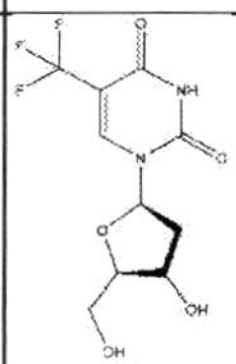 | Thymidine analogue Antimetabolite | Trifluridine exhibits anti-cancer properties by penetrating cancer cells and undergoing rapid conversion by thymidine kinase to its active monophosphate state. This is followed by subsequent phosphorylation to generate trifluridine triphosphate, which is readily incorporated into the DNA of tumour cells in place of thymidine bases. This disrupts DNA function, synthesis, and the proliferation of tumour cells. Additionally, trifluridine monophosphate also temporarily hinders thymidylate synthetase, a crucial enzyme for DNA synthesis that has been identified as upregulated in various cancer cell lines. Following phosphorylation by thymidine kinase, trifluridine transforms into its active monophosphate state within cancer cells. Subsequent phosphorylation yields trifluridine triphosphate, which effectively replaces thymidine bases in tumour cell DNA, disrupting DNA function and impeding tumour cell proliferation. Trifluridine monophosphate also transiently inhibits thymidylate synthetase. | Colorectal Cancer, Gastroesophageal junction adenocarcinoma |

**Table 1.** (Continued)

| S.No. | Name of drug candidate | Structure | Pharmacological class | Mechanism of inhibition | Application |
|---|---|---|---|---|---|
| 5. | Erlotinib | 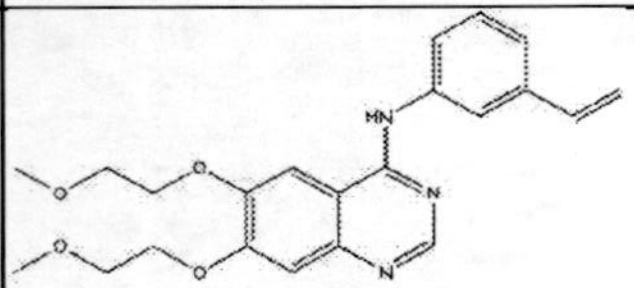 | Kinase (tyrosine) Inhibitor | Erlotinib works as an Epidermal Growth Factor Receptor (EGFR) Inhibitor. It does so by inhibiting the synthesis of Epidermal Growth Factor (EGF) by binding to Tyrosine Kinase and preventing its Intracellular Phosphorylation in the cell signalling pathway. EGF is one the main factors dominating cell proliferation and differentiation in the body. Hence, its depletion directly impacts cell growth and division in cancer. Erlotinib is acts against both normal and mutated EGFR. | Non-small cell Lung Cancer (NSCLC), Pancreatic Cancer, Oesophageal Squamous Cell Cancer (ESCC), Oesophageal Adenocarcinoma |
| 6. | Gefitinib | 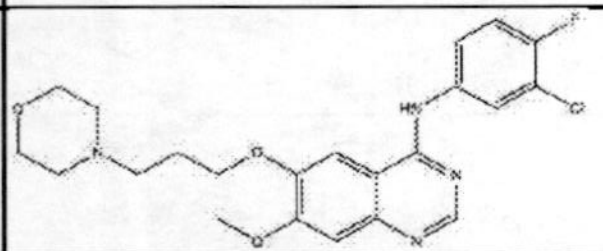 | Kinase (tyrosine) Inhibitor | Similar to erlotinib. | Non-small cell Lung Cancer (NSCLC) |
| 7. | Tegafur | 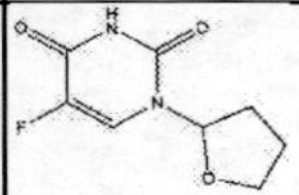 | Antimetabolite | Following administration, tegafur is converted into 5-FU, an active antineoplastic metabolite. In tumor cells, 5-FU undergoes phosphorylation to form active anabolites such as FdUMP. The binding of FdUMP and reduced folate to thymidylate synthase results in the formation of a ternary complex that inhibits DNA synthesis. Furthermore, FUTP is also incorporated into RNA, disrupting its normal functions. | Gastric cancer, Colorectal cancer |

| S.No. | Name of drug candidate | Structure | Pharmacological class | Mechanism of inhibition | Application |
|---|---|---|---|---|---|
| 8. | Thioguanine | 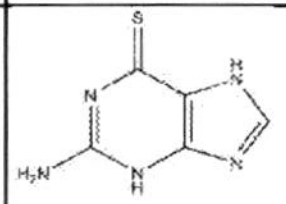 | Antimetabolite | Thioguanine is a purine analogue. It works against cancer cell growth by acting as a competitive inhibitor for several products in the *de novo* purine synthesis. It competes against guanine and hypoxanthine for the enzyme hypoxanthine-guanine phosphoribosyl transferase (HGPRTase) and gets converted to 6-thioguanilic acid (TGMP) in the process. TGMP triggers the pseudo-feedback inhibition of glutamine-5-phosphoribosylpyrophosphate amido transferase. TGMP also competes for IMP dehydrogenase and prevents the conversion of Inosinic acid (IMP) to Xanthylic acid (XMP). | Prostate Cancer, Fibrosarcoma, Gastric Cancer, Pancreatic Ductal Adenocarcinoma (PDAC) |

In the context of anticancer drug discovery, heterocyclic compounds are crucial to medical chemistry and synthetic organic research because of their immense biological significance (Hou and Xu 2022; Abbas et al., 2021). Because of its extensive pharmacological and biological effects, heterocycles containing nitrogen are actively working to catch the attention of scientists and researchers. Pyrimidine is one of the most pertinent limited templates in drug discovery (Abbas et al., 2021). Specifically, pyrimidine, 1,3-diazine, serves as the foundation for both RNA and DNA. Moreover, numerous biological effects, such as antiviral (Slagman and Fessner 2021), antibacterial (Ahmed, Choudhary, and Saleem 2023), antileishmanial (Poletto et al., 2021; Sunduru et al., 2006), anti-inflammatory (Rashid et al., 2021), neuroprotective (Manzoor et al., 2023), and cardiovascular (Irshad, Khan, and Iqbal 2021), are exhibited by pyrimidine and its derivatives as shown in Figure 1.

Hence, pyrimidine derivatives have gained significant attention in the field of medicinal chemistry due to their broad range of biological and pharmacological activities. A variety of pyrimidines, as well as pyrimidine-fused heterocyclic compounds, have been discovered to possess anticancer properties through various mechanisms and targets. Further, a number of molecules haves been authorised by US FDA like Gemcitabine, Vandetanib, Gefitinib, Pazopatinib, Lapatinib, Imatinib, Dasatinib, Nilotinib, Uramustine, Tegafur, Cytarabine, Methotrexate, and 5-Fluorouracil which are anticancer drugs that have pyrimidine scaffolds (Table 1). This indicates that medicinal chemists have played a key role in this field for several decades, employing various approaches. Among these, the investigation and discovery of pyrimidine derivatives or analogues have been particularly significant as anticancer agents. These compounds selectively target enzymes, proteins, or underlying pathways involved in oncogenesis (Lang et al., 2020; J. Zhang, Yang, and Gray 2009; Sun et al., 2020).

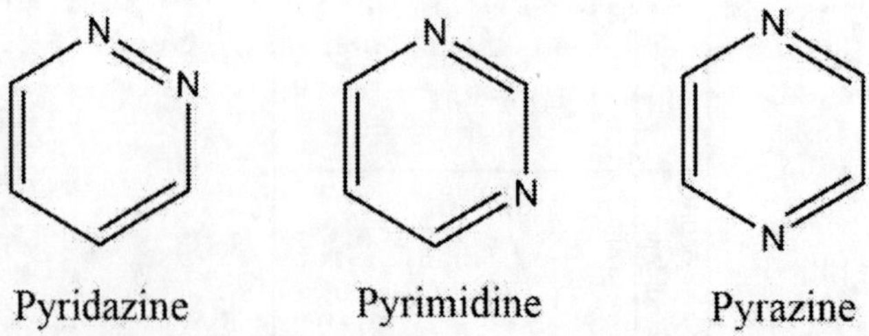
N
N
Pyridazine
N
N
Pyrimidine
N
N
Pyrazine

**Figure 2.** Chemical structures of diazines.

To summarize, this chapter functions as a comprehensive academic compilation, consolidating the accumulated insights from various research studies to provide a more nuanced outlook on the potential significance of pyrimidine derivatives in the continuous scientific endeavour to combat cancer.

## Pyrimidine: 1,3-diazine

Pyrimidine can be characterized as a basic aromatic heterocyclic moiety with an organic nature that bears a striking resemblance to pyridine. It is recognized as one of the three diazines (Figure 2), which are six-membered heterocyclics that contain two nitrogen atoms within the ring structure. It is a member of the electron-rich nitrogen heterocycle family.

Furthermore, within nucleic acids i.e., RNA and DNA, there exist three derivatives of pyrimidine, namely uracil, thymine, and cytosine (Shabarova and Bogdanov 2008; Moffatt and Ashihara 2002; Natarajan et al., 2023). Cytosine is a pyrimidine base that exists in both DNA and RNA. Its chemical composition comprises a heterocyclic aromatic ring, encompassing carbon and nitrogen atoms. This ring structure, consisting of six members, incorporates two nitrogen atoms and one oxygen atom. One of the nitrogen atoms is a component of an amine group (NH2), while the other nitrogen atom participates in a keto group (C = O). Additionally, cytosine possesses hydrogen atoms that are connected to the carbon framework. The disposition of these atoms contributes to the base's capacity to selectively pair with guanine through the formation of three hydrogen bonds in DNA, thus serving as a fundamental building block in the DNA double helix.

Thymine, exclusively present in DNA, bears a structural resemblance to cytosine but includes an additional methyl group. Its chemical structure also encompasses a six-membered heterocyclic ring containing two nitrogen atoms and one oxygen atom. Similar to cytosine, thymine possesses an amine group (NH2) and a keto group (C = O). Nonetheless, the distinguishing characteristic of thymine lies in the existence of a methyl group (-CH3) attached to the carbon ring structure. This unique methyl group sets thymine apart from cytosine and facilitates its specific base pairing with adenine in DNA through the establishment of two hydrogen bonds. This complementary base pairing is of paramount importance in maintaining the stability and integrity of the DNA molecule during the processes of replication and transcription.

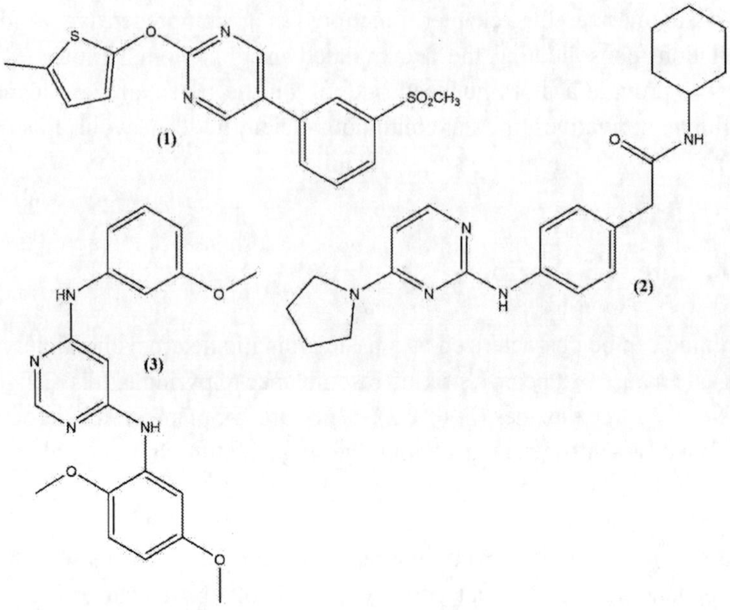
(1)
(2)
(3)
SO2CH3
NH
HN
NH

**Figure 3.** Chemical structures of di-substituted pyrimidine derivatives.

Uracil, primarily found in RNA, is a pyrimidine base. Its structure is similar to cytosine, but it lacks the methyl group present in thymine. Uracil consists of a six-membered heterocyclic ring, incorporating two nitrogen atoms and one oxygen atom, analogous to the structures of cytosine and thymine. It possesses an amine group (NH2) and a keto group (C = O), but it does not exhibit the methyl group present in thymine. The role of uracil in RNA entails specific base pairing with adenine through two hydrogen bonds, thereby contributing to the processes of genetic information transfer and protein synthesis within cells.

Hence, efforts have also been made in the way to mimic these biological molecules which can act as anticancer drugs. This can also be observed in the variety of currently used drugs as mentioned in previous section (Table 1).

## Structure Activity Relationship of substituted Pyrimidine Derivatives

The examination of Structure Activity Relationships (SAR) provides valuable insights into the molecular properties that contribute to receptor affinity and selectivity (Kerns and Di 2003; Kumari and Singh 2020). The compounds' promising nature can be attributed to the substitutions made in the hydrophobic domain. These compounds possess electron withdrawing and donating groups located at the ortho, meta, and para positions of the hydrophobic aryl ring. In general, it was observed that the substituted derivatives exhibited higher activity compared to the other derivatives. This can be explained by the fact that the substituted derivatives are better suited to fit into the receptor site. Pyrimidine's synthetic flexibility enables the creation of structurally varied derivatives, such as analogues produced by aryl ring replacement, pyrimidine nitrogen derivatization, and substitutions at of carbon at positions 2, 4, 5, and 6 (Selvam et al., 2015). Further, the efforts have been made to design of fused pyrimidine derivative in search of novel, safe, potent and effective anticancer agents (Nemr and AboulMagd 2020; Sherif and Yossef 2015; Abdellatif and Bakr 2021). In the following section, discussion on structure activity relationship such fused or substituted pyrimidine derivatives with respect to their anticancer activity.

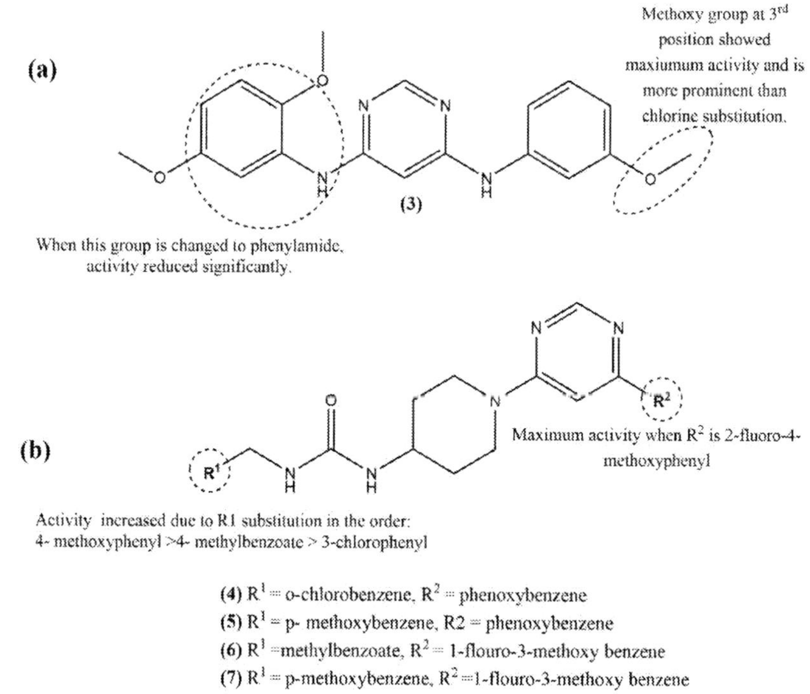
(a)
Methoxy group at 3rd position showed maxiumum activity and is more prominent than chlorine substitution.
(3)
When this group is changed to phenylamide, activity reduced significantly.
(b)
Maximum activity when R2 is 2-fluoro-4-methoxyphenyl
Activity increased due to R1 substitution in the order: 4- methoxyphenyl >4- methylbenzoate > 3-chlorophenyl
(4) R1 = o-chlorobenzene, R2 = phenoxybenzene
(5) R1 = p- methoxybenzene, R2 = phenoxybenzene
(6) R1 =methylbenzoate, R2 = 1-flouro-3-methoxy benzene
(7) R1 = p-methoxybenzene, R2 =1-flouro-3-methoxy benzene

**Figure 4.** SAR analysis of discovered 4, 6-disubstituted pyrimidine compounds (a) (Y. Zhang et al., 2018) (b) (Mule et al., 2016).

## Di-Substituted Pyrimidines

Reddy et al., synthesised 2,5-disubstituted pyrimidines by Suzuki coupling reaction with good yields and showed moderate anticancer effectiveness against HeLa cell lines using the MTT cell proliferation assay. Compound **1** (Figure 3) containing (5-methylthiophen-2-yl) methoxy at 2$^{nd}$ position showed the highest activity, with an $IC_{50}$ of 82.7 μM (Reddy et al., 2015). It was observed that different substitutions at 2$^{nd}$ and 5$^{th}$ positions resulted in decrease in activity. Further substitution at this place also results in decrease in activity. Aurora kinases are considered to play a crucial role in regulating meiosis and its overexpression has been reported as a cause of cancer progression. Hence, it stand as an important target in search of new anticancer therapies (Tang et al., 2017). Xu et. al. designed, synthesied and evaluated a series of novel Aurora Kinase inhibitors. The MTT assay was used to assess a few novel 2,4-disubstituted pyrimidines for their anti-proliferative properties. Compound **2** (Figure 3) demonstrated significant aurora kinase inhibition against both aurora A and B kinase and was moderately to highly active against the A549 ($IC_{50}$ = 12.05 ± 0.45 μM), HTC-116 ($IC_{50}$ = 1.31 ± 0.41 μM), and MCF-7 ($IC_{50}$ = 20.53 ± 6.13 μM) cell lines. Upregulation of Bax and downregulation of Bcl-xl caused apoptosis. According to SAR, replacing the benzene ring with a cyclohexyl group resulted in increased activity, but replacing the NH in urea with CH2 resulted in a decrease in activity (Xu et al., 2020).

Zhang and his co-worker designed a series a N4,N6-disubstituted pyrimidine 4,6 diamine derivatives inhibiting EGFR for treatment of non-small cell lung cancer (NSCLC) (Y. Zhang et al., 2018). When 3-methoxy group was placed at N4 and various substitutions were made at N6, it was observed that compound **3** (Figure 3) showed the most potent action and highest activity and SAR was reported as shown in Figure 4 (a). The *in-vitro* effects of a number of new 4, 6-disubstituted pyrimidine derivatives prepared by Mule and his coworkers on the malignant cell lines SIHA, IMR-32, A549, PANC-1, DU145, and MDA-MB-231 were assessed. It was discovered that compound **4** was an effective inhibitor of IMR32, compound **5** of MDA-MB-231, compound **6** of SIHA and DU145, and compound **7** of PANC-1 and A549, respectively (Figure 4(b)) (Mule et al., 2016). Figure 4 (b) shows the structure activity relationship of the synthesised compounds.

## Tri-substituted Pyrimidines

Kraljević et al., synthesised C-5 substituted pyrimidine using 5- Fluorouracil as a positive control. The compound **8** (5-chloroethyl 2,6 dichloro pyrimidine) ($IC_{50} = 0.8 \pm 0.2$ μM) caused DNA damage and a cytostatic effect on the HCT-116 cancer cell line that caused DNA damage and cell cycle arrest at the G2/M phase. Based on structural and molecular recognition, the compound with the highest potential was found to contain two aromatic and one aliphatic chlorine atoms connected to the pyrimidine ring (Figure 5) (Kraljević et al., 2012).

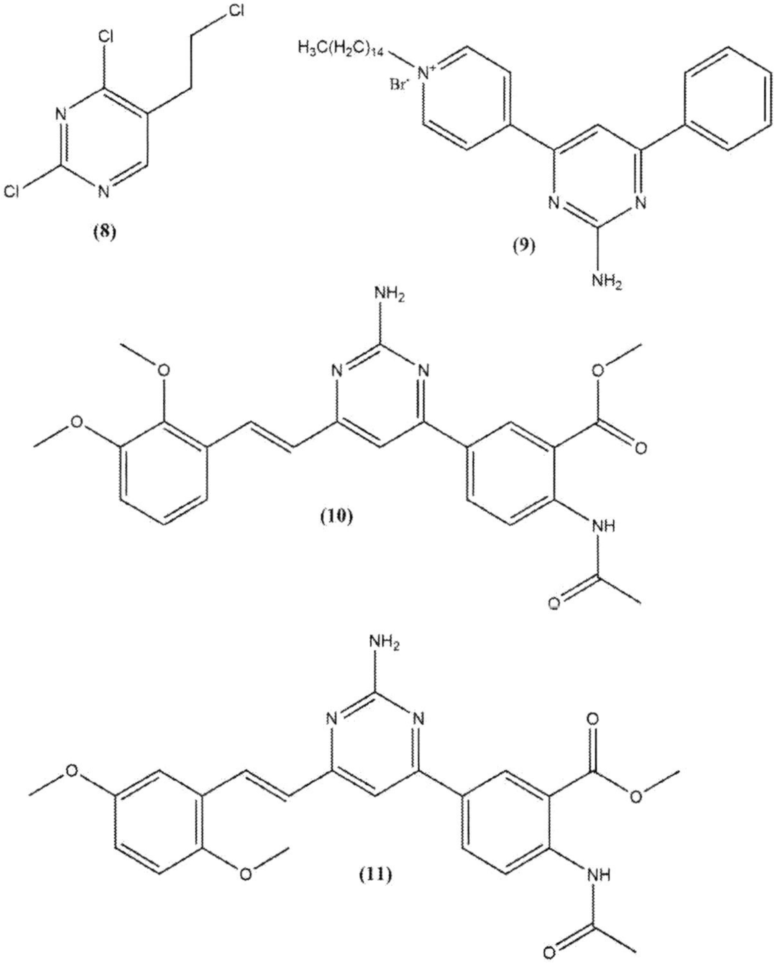
(8)
(9)
(10)
(11)

**Figure 5.** Chemical structures of tri-substituted pyrimidine derivatives.

Kahriman et al., prepared a new series 2,4,6-trisubstituted pyrimidines and their *N*-alkyl bromide derivatives based upon methoxy substituted azachalcones as the starting materials. The compound **9** (Figure 4) was designed by the reaction which goes on either by the 1,2-addition and/or 1,4-addition (Michael addition) of guanidine to the *α*, *β*-unsaturated carbonyl part of azachalcone, traced by cyclization to give the corresponding 2-amino-4,6-disubstituted pyrimidine compound. There was a remarkable activity from this compound against A549,Hep3B,HT29 FL, MCF-7 and HeLa cell with an $IC_{50}$ range of 2 to 10 μm/ml (Kahriman et al., 2019). Cheremnykh et al., used one pot multicomponent approach to synthesize a series of 2,4,6-trisustituted pyrmidine with an antharanilic acid ester moiety from methyl 5-(ethynyl)anthranilate, aroyl or cinnamoyl chlorides, and various amidines. Using standard MTT assay. The cytotoxic activity of each compound was assessed in relation to model cancer cell lines (CEM-13, U-937, MDA-MB-231, BT-474, DU-145). These compounds **10** and **11** (Figure 5) were recognized as the most active in the group and as CDK9 inhibitors. According to SAR studies, the main sources of activity are the methyl at position R2, the (E)-styryl moiety at position C-6, and the presence of a methyl anthranilate moiety with an EDG at position C-4, which results in increased activity (Cheremnykh et al., 2019).

## Tetra-Substituted Derivatives

Mohsen et al., synthesised a novel class of 2,4-disubstituted-2-thopyrimidine derivatives by synthesizing a series of 2- thioxopyrimidinones through one pot reaction of thiourea and ethyl cyanoacetate with alkylbenzaldehyde or alkoxybenzaldehyde in the presence of anhydrous $K_2CO_3$ to yield certain intermediates which are further reacted with with different 2-bromoacetophenones to give the final product (Abdel-Mohsen et al., 2019). The synthesised compounds inhibited VEGFR-2 when tested on the cancer cell lines of HepG2 and UO-31. Compound **12** ($IC_{50}$ = 1.23 μM) and compound **13** ($IC_{50}$ = 3.78 μM) were found to be active inhibitor of VEGFR-2 as well as displayed activity against HePG2 with the $IC_{50}$ = 13.06 μM and $IC_{50}$ = 8.35 μM respectively. As per the SAR studies, the hydrophobic interaction between the phenyl group at position 4 and the substitution at positions 2 of the thiouracil moiety was responsible for the potency (Figure 6).

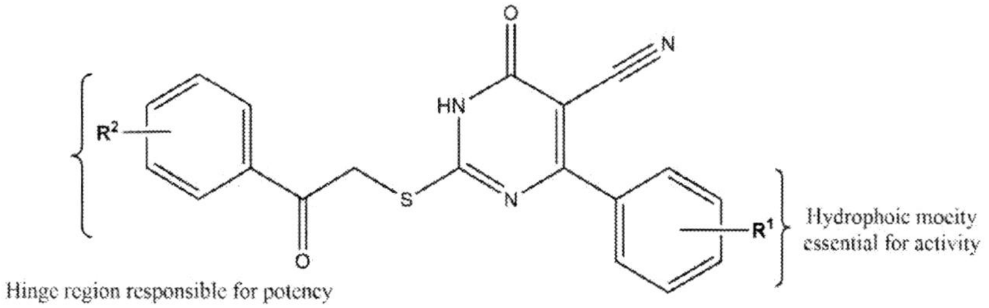
O
N
HN
R2
S
N
O
R1
Hydrophoic mocity essential for activity
Hinge region responsible for potency

**(12)** $R^1$ = 4-$OCH_3$, $R^2$ = 4-$OCH_3$
**(13)** $R^1$ =4-$OCH_3$, R2 = 2-$OCH_3$

**Figure 6.** Chemical structures and SAR of 2,4,5,6-tetra-substituted pyrimidine derivatives.

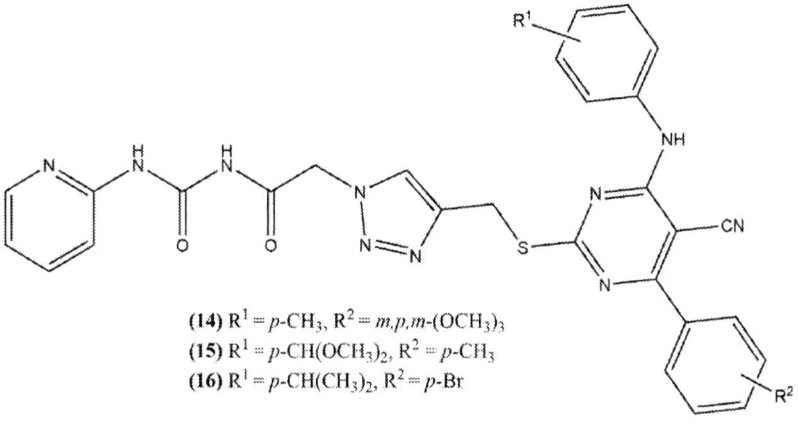
R1
NH
N
H
N
H
N
O
O
N
N
N
S
N
CN
N
R2

**(14)** $R^1$ = *p*-$CH_3$, $R^2$ = *m,p,m*-$(OCH_3)_3$
**(15)** $R^1$ = *p*-$CH(OCH_3)_2$, $R^2$ = *p*-$CH_3$
**(16)** $R^1$ = *p*-$CH(CH_3)_2$, $R^2$ = *p*-Br

**Figure 7.** Chemical structure of compounds **14**, **15** and **16** (2,4,5,6-tetra-substituted pyrimidine derivatives).

Ma et al., synthesised a novel series 1,2,3-triazole–pyrimidine–urea hybrids as potential anticancer agents by prolonged heating of aldehydes, ethylcyanoacetate, and thiourea in ethanol, in the presence of potassium carbonate (Ma et al., 2015). The intermediates were allowed to react with propargyl bromide in the presence of phosphorous oxychloride, in dioxane to yield the target derivatives. These target derivatives underwent click reaction to give some highly reactive intermediate which were finally reacted with different arylamines to obtain compounds the target compounds. They showed their anticancer activity against MGC-803, B16- F10, EC-109, and MCF-7 cancer cell lines and it was observed that the compounds 1**4, 15** and **16** had strong anti- B16-F10 activity with an IC50 = 32 nM, 35 nM, and 42 nM respectively (Figure 7). The SAR showed that the presence of 4-methyl and 4-methoxyl attributes at $R^1$ gives better activity than 2-methyl, 3-methyl, and 2-methoxyl substitutions, and that the electronic effect on the phenyl ring affects the activity, giving an electron donating group more activity.

## Pyrazolopyrimidines

In 2006, Li et al., synthesized compound **17**, which showed potent antitumor activity in vitro against the Bel-7402 liver cancer cell line and HT-1080 fibrosarcoma cell line (Figure 8). (Li et al., 2006). In 2010, Abdel-Aziz et al., described a simple method for synthesizing a series of pyrazolo[1,5-a]pyrimidines derivatives, which showed promising antitumor effects against colon cancer (CaCo-2) and normal fibroblast (BHK) cells in laboratory studies. The key starting material, E-3-(N,N-dimethylamino)-1-(3-methylthiazolo[3,2-a]benzimidazol-2-yl)prop-2-en-1-one, was made in good yield using Gold's reagent. Treating this enaminone compound with 5-amino-1*H*-pyrazole yielded pyrazolo[1,5-a]pyrimidines. The structure of pyrazolo[1,5-a]pyrimidine was confirmed by X-ray diffraction. The synthesized compounds were evaluated for antitumor effects on colon CaCo-2 cancer cells and cytotoxicity against normal BHK fibroblasts. Some of the tested compounds showed inhibitory effects on cell growth. One compound **18** displayed significant antitumor activity against CaCo-2 cells ($IC_{50}$ = 0.5 μg/mL) along with lower toxicity in BHK cells ($IC_{50}$ = 2.3 μg/mL) (Figure 8) (Abdel-Aziz, Saleh, and El-Zahabi 2010).

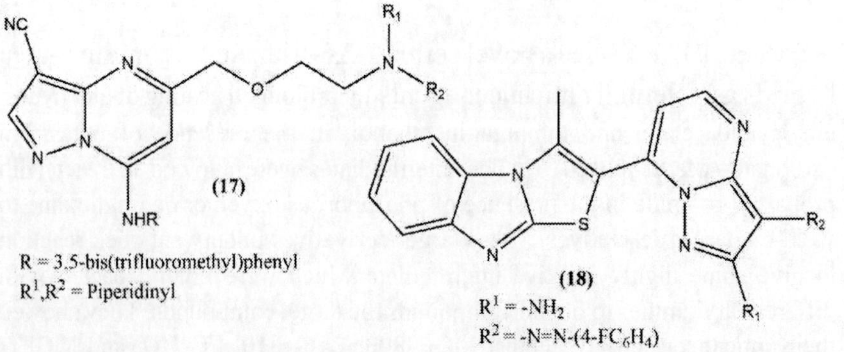
NC
N
N
N
O
R1
N
R2
NHR
(17)
R = 3,5-bis(trifluoromethyl)phenyl
R1,R2 = Piperidinyl
N
N
S
N
N
N
R2
R1
(18)
R1 = -NH2
R2 = -N=N-(4-FC6H4)

**Figure 8.** Chemical structures of compounds **17** and **18**.

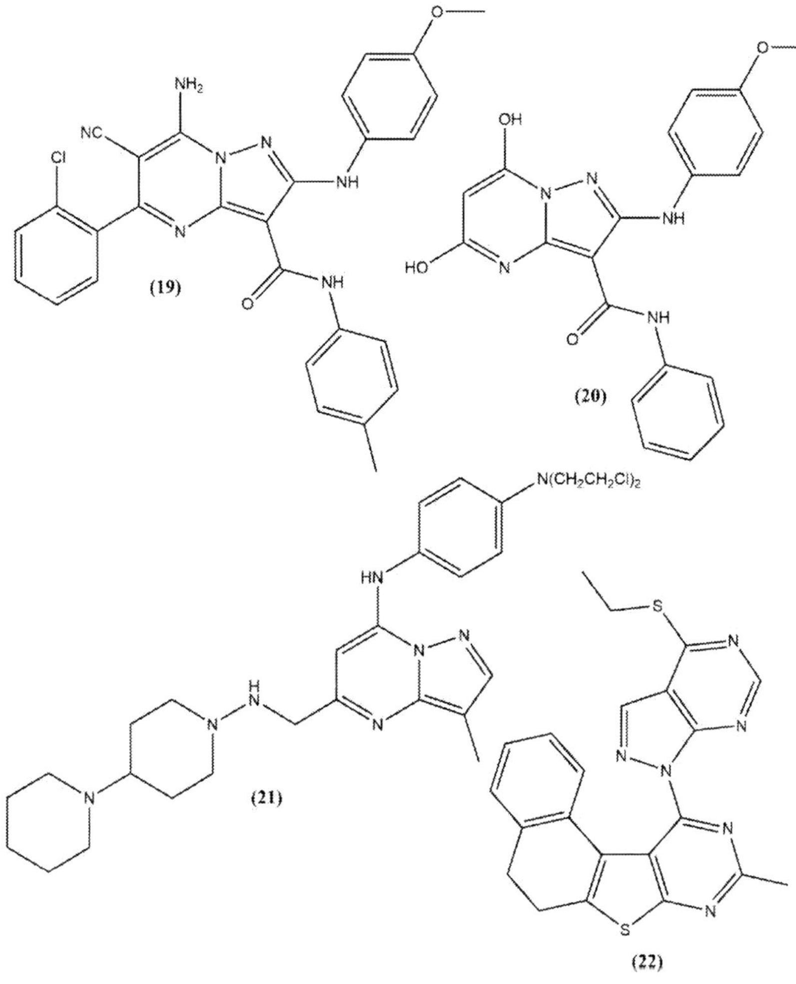
NH2
NC
Cl
N
NH
O
(19)
OH
HO
(20)
N(CH2CH2Cl)2
HN
(21)
S
(22)

**Figure 9.** Chemical structures of compounds **19-22.**

Hassan et. al. performed MTT assay to assess the cytotoxicity of synthesised pyrazolo [1,5-a] pyrimidine derivatives against PC-3, HCT116, and HepG-2 cancer cell lines. The results showed that **19** ($IC_{50}$ = 67.27 ± 3.8 μM/mL) and **20** ($IC_{50}$ = 58.44 ± 3.8 μM/mL) had the highest activity against PC-3 and HCT116 cell lines respectively (Figure 9) (Hassan et al., 2017). According to SAR tests, chlorine atoms at position 2 was more active than those at positions 3 and 4, and the order of antitumor activity against the cell lines was 4-methylphenyl > 4-chlorophenyl > phenyl derivative. Zhao et al., also used the MTT assay to evaluate the anticancer activity of the pyrazolo [1,

5-a] pyrimidine derivatives containing nitrogen mustard moiety against the cell lines A549, SH-SY5Y, HepG2, MCF-7, and DU145. Compound **21** caused apoptosis in each of the five cancer cell lines, inhibiting proliferation at the G1 phase of the cell cycle ($IC_{50}$ range: 0.2 to 8.3 μM) (Figure 9). The pyrrazolo-pyrimidine derivatives with N-mustard pharmacophore at C-7 and another substituent at C-5 showed strong in vitro cytotoxicity; however, the molecule generated when N-mustard pharmacophore is attached at C-5 and different aniline moieties at C-7 is found to be ineffective (Zhao et al., 2016). Among the series of substituted pyrazolo [3, 4-d] pyrimidines synthesised by Rashad et. al., compound **22** (Figure 9) was shown to be the most active when evaluated *in-vitro* against MCF-7 cancer cell lines (Rashad, Mahmoud, and Ali 2011). The SAR was also analysed as shown in Figure 10.

Mishra and his colleagues designed a novel series of of urea and pyrazolo[3,4-d]pyrimidine hybrids. Compounds **23** through **27** among them demonstrated encouraging cytotoxicity against tested cancer cell types (Figure 11). When compared to doxorubicin, compound **27** was the most effective derivative and showed superior cytotoxicity against all tested cell types. Compound 5 effectively stopped the advancement of the cell cycle and showed excellent apoptosis in A549 cells. Furthermore, it exhibited tumoricidal activities in vivo in lung adenocarcinoma model of xenograft naked mice.

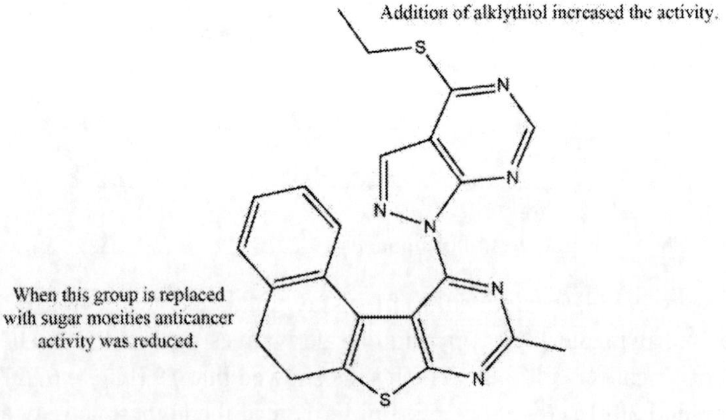
Addition of alklythiol increased the activity.
S
N
N
N
N
N
When this group is replaced with sugar moeities anticancer activity was reduced.
S
N

**(22)**

**Figure 10.** SAR examination of compound **22**.

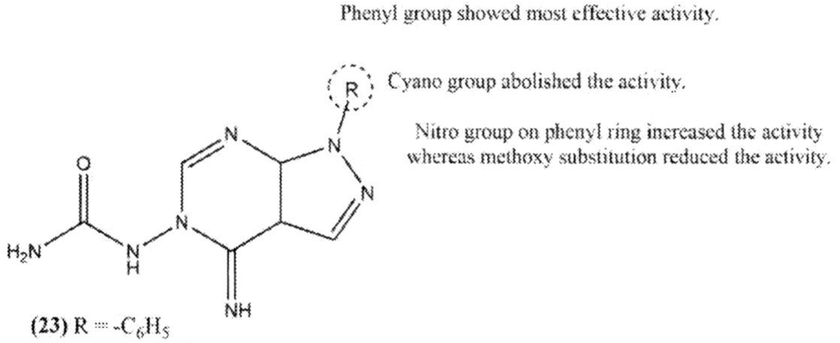
Phenyl group showed most effective activity.
Cyano group abolished the activity.
Nitro group on phenyl ring increased the activity whereas methoxy substitution reduced the activity.

(23) R = -$C_6H_5$
(24) R = *p*-F-$C_6H_5$
(25) R = *m*-Br-$C_6H$
(26) R = *m*-$NO_2$-$C_6H_5$
(27) R = *p*-$NO_2$-$C_6H_5$

**Figure 11.** Chemical structures and SAR of compounds **23-27**.

Compound **28** was found to be the most effective inhibitor of both CDK2 ($IC_{50}$ = 21 nM) and CDK5 ($IC_{50}$ = 35 nM) after a series of 5-substituted 3-isopropyl-7-[4-(2-pyridyl)benzyl] amino-1(2)*H*-pyrazolo[4,3-d]pyrimidine derivatives were reported in a study (Vymětalová et al., 2016). It was noted that the hydroxyalkylamines located at the pyrazolo-pyrimidine ring's fifth position were primarily responsible for the action. In a different investigation, 1H-pyrazolo [4,3-d] pyrimidin-7(6*H*)-ones and their anticancer efficacy were described employing microwave-assisted synthesis. Using the MTT assay, the compounds were evaluated against human cancer cell lines HeLa, CAKI-I, PC-3, MiaPaca-2, and A549. Among all the synthesised compounds, compound **29** demonstrated the highest efficacy against all cancer cell lines with an $IC_{50}$ range of 14 to 38 μM, an apoptotic mechanism, and mTOR inhibition at nanomolar potency.

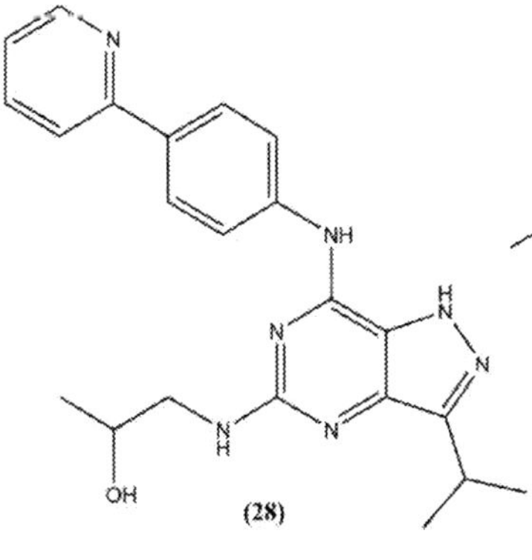

(28)

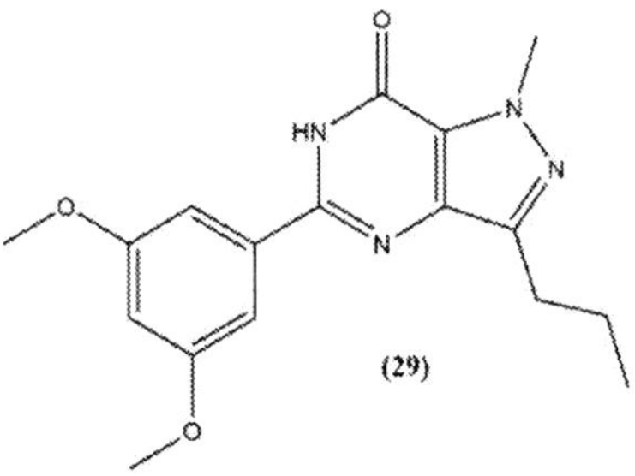

(29)

**Figure 12.** Chemical structures of compounds **28** and **29**.

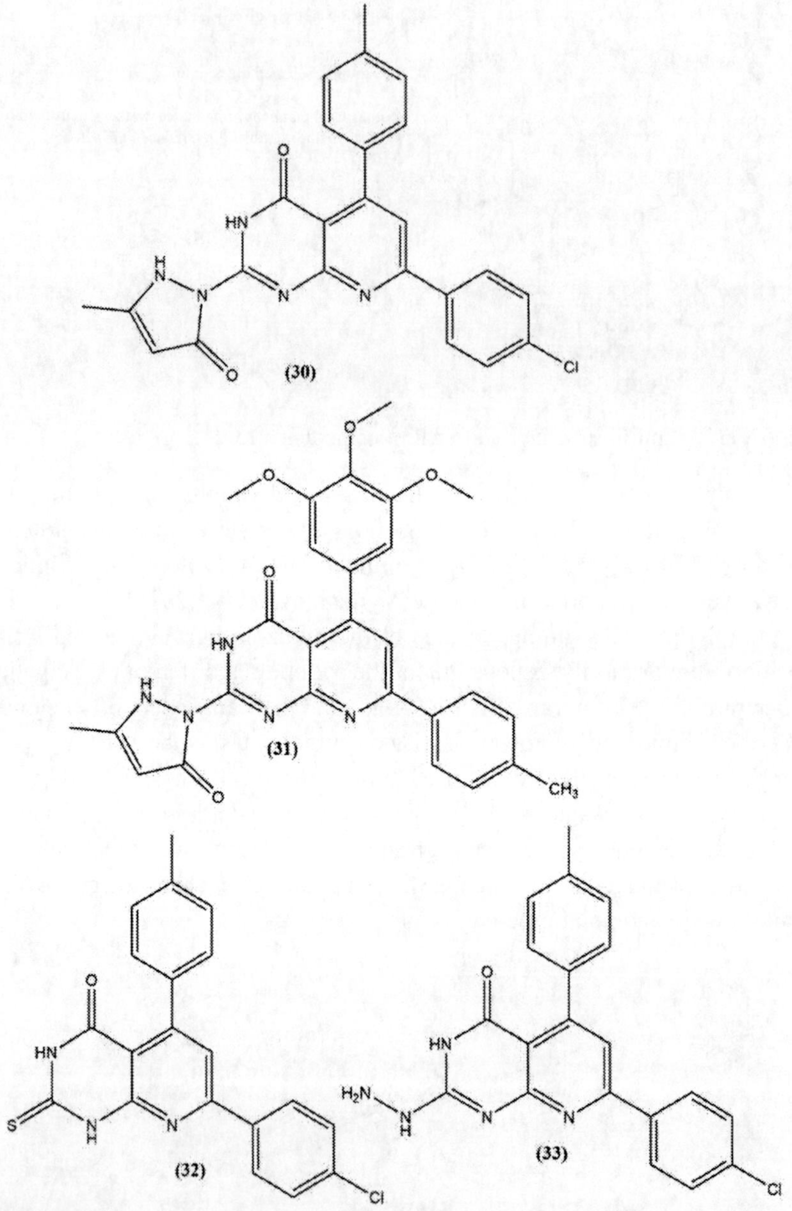
(30)
(31)
CH3
(32)
(33)
Cl

**Figure 13.** Chemical structure of pyridopyrimidine derivatives (30-33).

## Pyridopyrimidines

In 2018, Elzahabi et al., successfully synthesized pyrido[2,3-d]pyrimidine derivatives and tested their ability to inhibit the growth of five cancer cell lines. Pyrido[2,3-d]pyrimidines were found to significantly inhibit a variety of kinases, including tyrosine kinases, PI3K, and CDK4/6 (Elzahabi et al., 2018). In particular, compounds **30** and **31**, which had 4-methylphenyl and 4-chlorophenyl groups at positions 5 and 7, respectively, shared the same pyridopyrimidine scaffold (Figure 13). Converting the thioxo group in **32** to a hydrazide in **33** significantly enhanced the anti-hepatic cancer activity. The electronic properties were able to positively impact anticancer activity due to the hydrophilic electron-rich nature of the hydrazide moiety. The synthesized compounds displayed significant growth inhibitory effects in HepG-2, HCT-116, and PC-3 cell lines compared to doxorubicin, but showed relatively low activity in MCF-7 and A-549 cancer cells. Furthermore, the potent anticancer molecule **30** inhibited PDGFR β, EGFR, and CDK4/Cyclin D1 kinases.

Zhang et al., reported the scaffold-hopping synthesis of ERK/PI3K dual inhibitors by substituting the 1*H*-pyrazolo[3,4-d]pyrimidine scaffold with pyrido[3,2-d]pyrimidine (L. Zhang et al., 2020). Compound **34** with the pyrido[3,2-d]pyrimidine scaffold showed suitable inhibitory activities against ERK2 and PI3Kα, while compound **35** with the pyrido[2,3-d]pyrimidine scaffold had 1.8% and 15.1% inhibitory activity at 1 μM concentration on ERK2 and PI3Kα, respectively. Compound **36**, a potent and very effective ERK and PI3K dual inhibitor, was discovered during initial SAR analysis. Compound **36** exhibited modest ERK and PI3K inhibitory and anti-proliferative properties. Although 74 had only acceptable pharmacokinetic profile in SD rats, with a moderate half-life ($t1/2 = 2.32$ hrs) after intravenous administration, it showed significant anticancer activity in vivo in an HCT-116 xenograft model without causing obvious adverse effects.

There was a report on new pyrido[3′,2′:4,5]furo[3,2-d]pyrimidines (Naresh Kumar et al., 2016). Using the MTT assay, the compounds were tested against Hela, neuro-2a, Colo 205, and A549 tumour cell lines. The most powerful compounds were **37** and **38**, whose $IC_{50}$ ranged from 5.8 to 3.6 μM. The pyridine ring system of furo [2, 3-b] was less active than the pyrrimidinone ring. Substitution of nitrogen on N3 led to a slight decrease in cytotoxicity (Figure 15). In an another study HeLa, A549, PANC 1, and MDA MB-231 cancer cell lines were used to investigate the anticancer potential of novel triazole/isoxazole functionalized 7-(trifluoromethyl)pyrido [2,3-

d]pyrimidine derivatives at concentrations less than 10 μM. The most effective compounds, **39** against PANC1 and **40** against A549 (Figure 15), outperformed the positive control nocodazole in terms of activity. The compound's strong anticancer activity was brought about by the lengthy alkyl/perfluoroalkyl chain at R on the triazole ring and the presence of an ethyl chain at C-2 position. Regardless of the triazole ring or C-2 position alterations, a number of compounds exhibited activity.

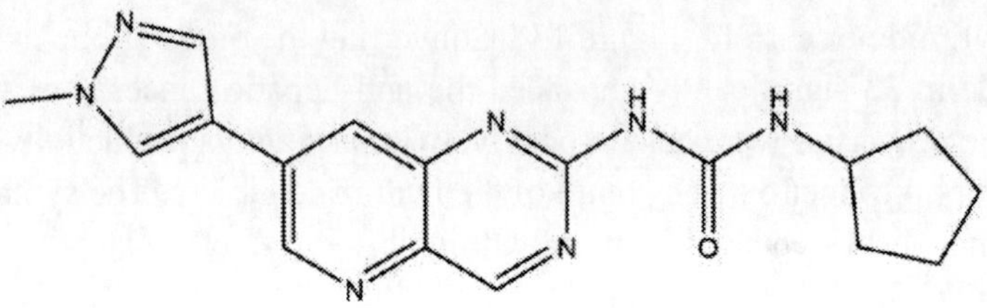

**(34)**

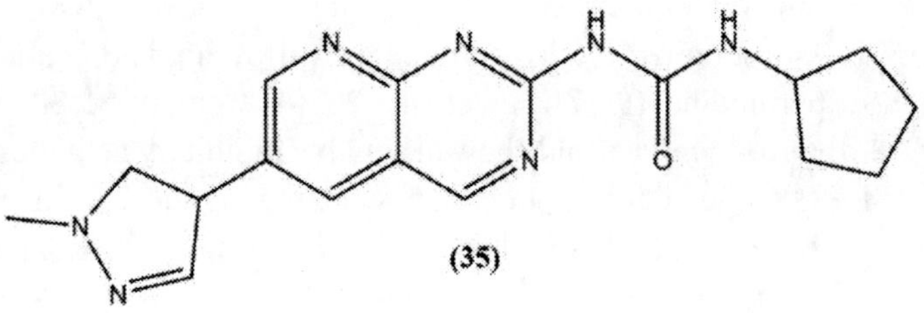

**(35)**

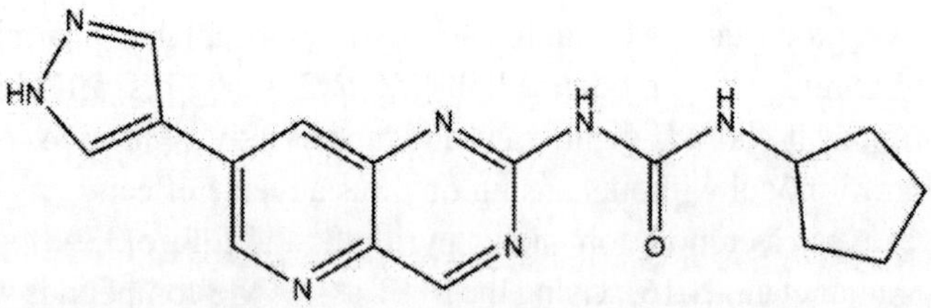

**(36)**

**Figure 14.** Chemical structure of Pyrido[2,3-*d*]pyrimidine and pyrido[3,2-*d*]pyrimidine derivatives as dual inhibitors of ERK2 and PI3Kα (**34-36**).

## Pyrrolopyrimidines

Pyrrolopyrimidine derivatives especially, pyrrolo[2,3-d] pyrimidine derivatives are popularly used in cancer therapy due to their structural similarities with cellular purines. Rania H. Abd El-Hameed et al., 2018 studied the anticancer nature of pyrrolopyrimidines. They focussed on pyrrolopyrimidine derivatives which acted as cyclin dependant kinase-2 (CDK-2) inhibitors and synthesised 35 pyrrolopyrimidine derivatives (Abd El-Hameed and Sayed 2018). Out of these five compounds showed average anticancer properties with compound **41** being most promising. They took pyrollopyrimidin-4-thiones as initial reactant. This was reacted with chloroethyl acetate to form associated esters (like compound **41**). These were further reacted with hydrazine hydrate to get acetohydrazide products and similarly novel pyrrolopyrimidine derivatives (like compound **42, 43, 44 and 45**) (Figure 16). They employed several cancer cell lines in their studies like- K-562 (Leukaemia), NCI-H522 (Non-small lung Cancer), U251 (CNS cancer), SK-MEL-2 (Melanoma), T-47D (Breast Cancer) etc. Compound 1 had maximum mean growth percentage (69.14%) out of all synthesised pyrrolopyrimdine derivatives. Compound **41** consisted of ethyl thioacetate substituent at 4$^{th}$ position, phenyl groups at both position five and six and p-methylphenyl at 7$^{th}$ position. These substituitions can be taken as prerequisites for moderate anticancer activity. Any changes made at these positions led to a comparable decrease in anticancer nature. Further hydrazide formation and cyclization also resulted in overall decrease in activity. Compound 1 showed average activity against 7 cell lines- Leukaemia cell lines;K-562 and RPMI-8226, CNS cell line; UT51, Melanoma cell line; SK-MEL-5, NSC-LC cell lines; HOP-92, EKVX, NCI-H522. It shows especially high anticancer activity against T-47D, breast cancer cell line. Further, in compounds **42-45** the formation of N-arylidine acetohydrazide caused a regain in moderate anticancer activity. Moreover, the compounds **42-44** show average anticancer activity against one cell line. Compound **45** shows moderate anticancer activity against 5 cancer cell lines namely, MCF-7 (Breast Cancer), MDA-MB-468 (Breast Cancer), A498 (Renal Cancer), HOP-92 (NSC-lung cancer), NCI-H226 (NSC-lung cancer).

In 2019, Kilic-Kurt et al., synthesised novel pyrrolopyrimidine derivatives consisting of pyrrolo[2,3-d] pyrimidine derivatives with urea moiety (Kilic-Kurt et al., 2019). They employed three cancer cell lines- A549, PC-3 and MCF-7. Most synthesised compounds showed higher anticancer property against A549 and PC3 cell lines when, Imatinib was taken as standard.

Promising $IC_{50}$ values of 0.35, 1.48 and 1.56μM were shown by some derivatives against A549 cell line. Some of the most promising derivatives were obtained by replacing the functional group at X position with a CF-2 group. Chloro-, Bromo- or fluoro- substitutions at this position at m- or p-positions did not increase cytotoxicity of the compound as shown in Figure 17. However, Cl substitution at Y position increased the anticancer ability across all cancer cell lines-A549, MCF-7 and PC3. Fluoro- substitution at Z position demonstrated highest anticancer property.

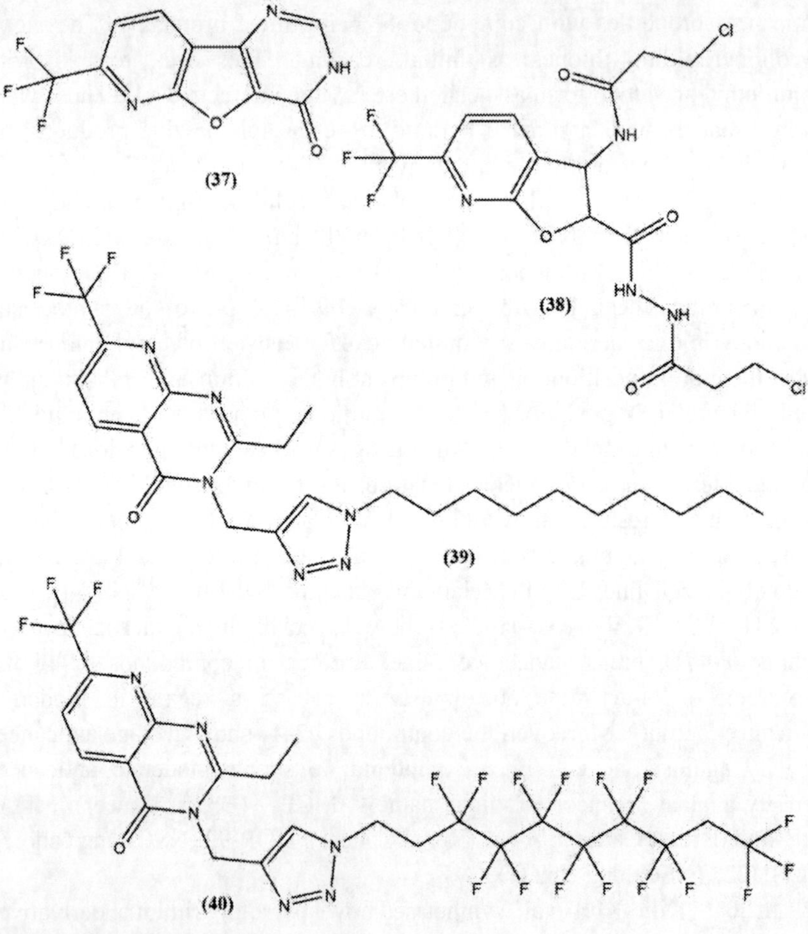
(37)
(38)
(39)
(40)

**Figure 15.** Chemical structures of compounds **37-40.**

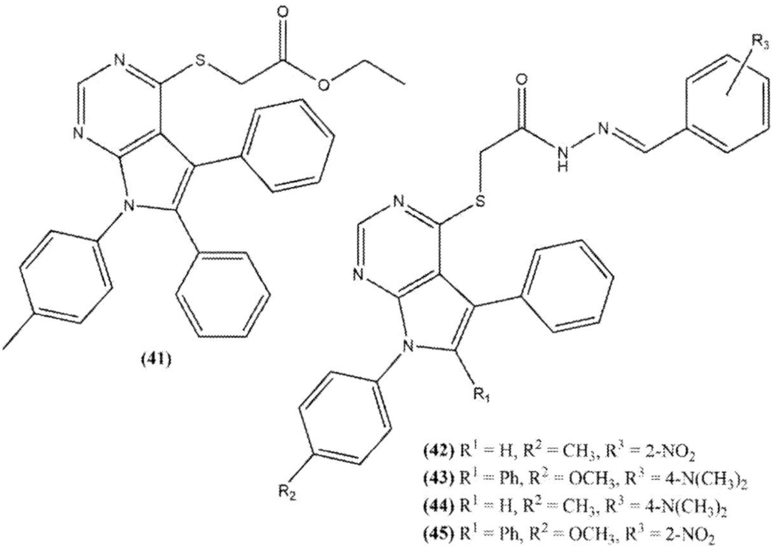
(41)
(42) R[1] = H, R[2] = $CH_3$, R[3] = 2-$NO_2$
(43) R[1] = Ph, R[2] = $OCH_3$, R[3] = 4-$N(CH_3)_2$
(44) R[1] = H, R[2] = $CH_3$, R[3] = 4-$N(CH_3)_2$
(45) R[1] = Ph, R[2] = $OCH_3$, R[3] = 2-$NO_2$

**Figure 16.** Chemical structures of compounds **41-45**.

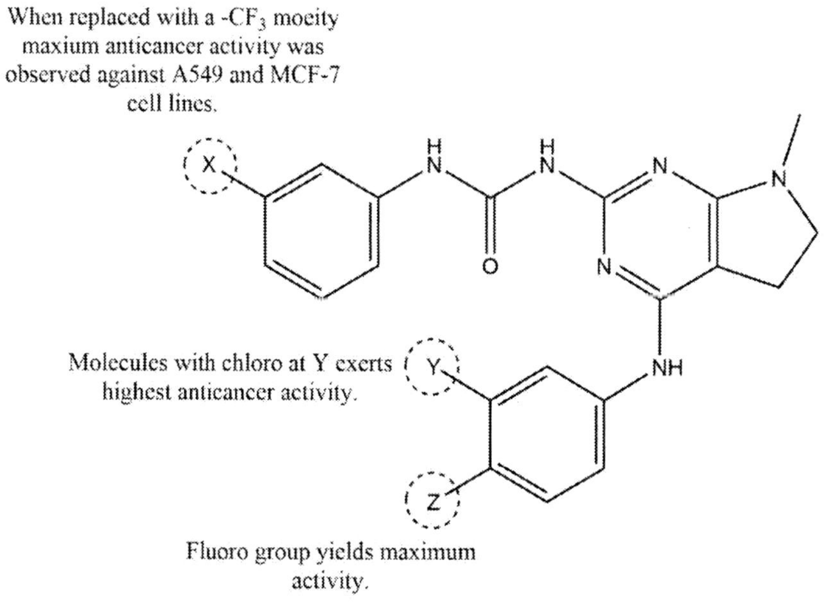
When replaced with a -$CF_3$ moeity maxium anticancer activity was observed against A549 and MCF-7 cell lines.
Molecules with chloro at Y exerts highest anticancer activity.
Fluoro group yields maximum activity.

**Figure 17.** SAR analysis of pyrrolo[2,3-d] pyrimidine derivatives with urea moiety (Kilic-Kurt et al., 2019).

## Conclusion

Since, pyrimidine is found in naturally occurring nucleotides, it is an important scaffold whose anticancer profile is currently being investigated in great detail. The anticancer potential of pyrimidine substituted at different locations and pyrimidine fused with other heterocyclic rings has been covered in this chapter. In medicinal chemistry and drug development, the pyrimidine ring and its fused derivatives, such as pyrazolo[3,4-d]pyrimidine, pyrido[2,3-d]pyrimidine, and quinazoline, have proven to be excellent target molecules. The way these scaffolds worked was by blocking vital enzymes that control cell division, migration, growth, and metabolism. The pyrimidine scaffold has gained significant attention from the pharmaceutical industry and academic researchers in the past, as evidenced by the substantial number of studies and patents published i.e., from 2009 to the present. It is anticipated that as a result of intense research, a large body of literature has amassed throughout time, and the discipline of pyrimidine chemistry is still developing. It would be intriguing to observe the development of pyrimidines as potentially useful medicinal compounds. By delving more into the underlying data, a number of other novel compounds with anti-proliferative and/or related biological potential could be achieved in a number of ways. Crucially, structural variation, molecular fusion of the backbone of the main bioactive chemical, and in silico approaches could all lead to the logical development and creation of new bioactives. Anticipating the addition of new compounds will necessitate additional research for biological activity assessments of modified pyrimidines for various cancer types whose treatments are challenging to cure. These two heterocycles' biological actions will help scientists plan, categorise, and implement new techniques for finding brand-new anticancer medications.

## References

Abbas, Nahid, P M Gurubasavaraja Swamy, Prasad Dhiwar, Shilpa Patel, and D Giles. 2021. "Development of Fused and Substituted Pyrimidine Derivatives as Potent Anticancer Agents (A Review)." *Pharmaceutical Chemistry Journal* 54 (12): 1215–26. doi: https://doi.org/10.1007/s11094-021-02346-8.

Abd El-Hameed, Rania H., and Amira I. Sayed. 2018. "Synthesis of Novel Pyrrolopyrimidine Derivatives as CDK2 Inhibitors." *Pharmacophore* 9 (5): 29–49.

Abdel-Mohsen, Heba T., Adel S. Girgis, Abeer E. E. Mahmoud, Mamdouh M. Ali, and Hoda I. El Diwani. 2019. "New 2,4-Disubstituted-2-Thiopyrimidines as VEGFR-2

Inhibitors: Design, Synthesis, and Biological Evaluation." *Archiv Der Pharmazie* 352 (11). John Wiley & Sons, Ltd: 1900089. doi: https://doi.org/10.1002/ardp.201900089.

Abdel-Aziz, Hatem A., Tamer S Saleh, and Heba S. A. El-Zahabi. 2010. "Facile Synthesis and In-vitro Antitumor Activity of Some Pyrazolo [3, 4-b] Pyridines and Pyrazolo [1, 5-a] Pyrimidines Linked to a Thiazolo [3, 2-a] Benzimidazole Moiety." *Archiv Der Pharmazie: An International Journal Pharmaceutical and Medicinal Chemistry* 343 (1). Wiley Online Library: 24–30.

Abdellatif, Khaled R. A., and Rania B. Bakr. 2021. "Pyrimidine and Fused Pyrimidine Derivatives as Promising Protein Kinase Inhibitors for Cancer Treatment." *Medicinal Chemistry Research* 30 (1): 31–49. doi: https://doi.org/10.1007/s00044-020-02656-8.

Ahmed, Kainat, M. Iqbal Choudhary, and Rahman Shah Zaib Saleem. 2023. "Heterocyclic Pyrimidine Derivatives as Promising Antibacterial Agents." *European Journal of Medicinal Chemistry*. Elsevier, 115701.

Cheremnykh, Kirill P., Victor A. Savelyev, Mikhail A. Pokrovskii, Dmitry S. Baev, Tatyana G. Tolstikova, Andrey G. Pokrovskii, and Elvira E. Shults. 2019. "Design, Synthesis, Cytotoxicity, and Molecular Modeling Study of 2,4,6-Trisubstituted Pyrimidines with Anthranilate Ester Moiety." *Medicinal Chemistry Research* 28 (4): 545–58. doi: https://doi.org/10.1007/s00044-019-02314-8.

Elzahabi, Heba S. A., Eman S. Nossier, Nagy M. Khalifa, Rania A. Alasfoury, and May A. El-Manawaty. 2018. "Anticancer Evaluation and Molecular Modeling of Multi-Targeted Kinase Inhibitors Based Pyrido[2,3-d]Pyrimidine Scaffold." *Journal of Enzyme Inhibition and Medicinal Chemistry* 33 (1). Taylor & Francis: 546–57. doi: https://doi.org/10.1080/14756366.2018.1437729.

Hanahan, Douglas, and Robert A. Weinberg. 2011. "Hallmarks of Cancer: The next Generation." *Cell* 144 (5). Elsevier Inc.: 646–74. doi: https://doi.org/10.1016/j.cell.2011.02.013.

Hassan, Ashraf S., Mohamed F. Mady, Hanem M. Awad, and Taghrid S. Hafez. 2017. "Synthesis and Antitumor Activity of Some New Pyrazolo[1,5-a]Pyrimidines." *Chinese Chemical Letters* 28 (2): 388–93. doi: https://doi.org/10.1016/j.cclet.2016.10.022.

Hou, Wei, and Hongtao Xu. 2022. "Incorporating Selenium into Heterocycles and Natural Products - From Chemical Properties to Pharmacological Activities." *Journal of Medicinal Chemistry* 65 (6). American Chemical Society: 4436–56. doi: https://doi.org/10.1021/acs.jmedchem.1c01859.

Irshad, Nadeem, Arif-ullah Khan, and Muhammad Shahid Iqbal. 2021. "Antihypertensive Potential of Selected Pyrimidine Derivatives: Explanation of Underlying Mechanistic Pathways." *Biomedicine & Pharmacotherapy* 139. Elsevier: 111567.

Kahriman, Nuran, Vildan Serdaroğlu, Kıvanç Peker, Ali Aydın, Asu Usta, Seda Fandaklı, and Nurettin Yaylı. 2019. "Synthesis and Biological Evaluation of New 2,4,6-Trisubstituted Pyrimidines and Their N-Alkyl Derivatives." *Bioorganic Chemistry* 83: 580–94. doi: https://doi.org/10.1016/j.bioorg.2018.10.068.

Kerns, Edward H., and Li Di. 2003. "Pharmaceutical Profiling in Drug Discovery." *Drug Discovery Today* 8 (7). Elsevier: 316–23.

Kilic-Kurt, Zühal, Filiz Bakar-Ates, Yeliz Aka, and Ozgur Kutuk. 2019. "Design, Synthesis and in Vitro Apoptotic Mechanism of Novel Pyrrolopyrimidine Derivatives." *Bioorganic Chemistry* 83: 511–19. doi: https://doi.org/10.1016/j.bioorg.2018.10.060.

Kraljević, Tatjana Gazivoda, Mateja Klika, Marijeta Kralj, Irena Martin-Kleiner, Stella Jurmanović, Astrid Milić, Jasna Padovan, and Silvana Raić-Malić. 2012. "Synthesis, Cytostatic Activity and ADME Properties of C-5 Substituted and N-Acyclic Pyrimidine Derivatives." *Bioorganic & Medicinal Chemistry Letters* 22 (1): 308–12. doi: https://doi.org/10.1016/j.bmcl.2011.11.009.

Kumari, Archana, and Rajesh K. Singh. 2020. "Morpholine as Ubiquitous Pharmacophore in Medicinal Chemistry: Deep Insight into the Structure-Activity Relationship (SAR)." *Bioorganic Chemistry* 96. Elsevier: 103578.

Lang, K Damanpreet, Rajwinder Kaur, Rashmi Arora, Balraj Saini, and Sandeep Arora. 2020. "Nitrogen-Containing Heterocycles as Anticancer Agents: An Overview." *Anti-Cancer Agents in Medicinal Chemistry*. doi: http://dx.doi.org/10.2174/1871520620666200705214917.

Li, Juan, Yan Fang Zhao, Xiang Lin Zhao, Xiao Ye Yuan, and Ping Gong. 2006. "Synthesis and Anti-tumor Activities of Novel Pyrazolo [1, 5-a] Pyrimidines." *Archiv Der Pharmazie: An International Journal Pharmaceutical and Medicinal Chemistry* 339 (11). Wiley Online Library: 593–97.

Liu, Ya-Ping, Can-Can Zheng, Yun-Na Huang, Ming-Liang He, Wen Wen Xu, and Bin Li. 2021. "Molecular Mechanisms of Chemo- and Radiotherapy Resistance and the Potential Implications for Cancer Treatment." *MedComm* 2 (3). China: 315–40. doi: https://doi.org/10.1002/mco2.55.

Ma, Li-Ying, Bo Wang, Lu-Ping Pang, Miao Zhang, Sai-Qi Wang, Yi-Chao Zheng, Kun-Peng Shao, Deng-Qi Xue, and Hong-Min Liu. 2015. "Design and Synthesis of Novel 1,2,3-Triazole–Pyrimidine–Urea Hybrids as Potential Anticancer Agents." *Bioorganic & Medicinal Chemistry Letters* 25 (5): 1124–28. doi: https://doi.org/10.1016/j.bmcl.2014.12.087.

Malone, Eoghan R., Marc Oliva, Peter J. B. Sabatini, Tracy L. Stockley, and Lillian L. Siu. 2020. "Molecular Profiling for Precision Cancer Therapies." *Genome Medicine* 12 (1): 8. doi: https://doi.org/10.1186/s13073-019-0703-1.

Manzoor, Shoaib, Daniyah A. Almarghalani, Antonisamy William James, Md Kausar Raza, Tasneem Kausar, Shahid M. Nayeem, Nasimul Hoda, and Zahoor A. Shah. 2023. "Synthesis and Pharmacological Evaluation of Novel Triazole-Pyrimidine Hybrids as Potential Neuroprotective and Anti-Neuroinflammatory Agents." *Pharmaceutical Research* 40 (1). United States: 167–85. doi: https://doi.org/10.1007/s11095-022-03429-1.

Mattiuzzi, Camilla, and Giuseppe Lippi. 2019. "Current Cancer Epidemiology." *Journal of Epidemiology and Global Health* 9 (4). Switzerland: 217–22. doi: https://doi.org/10.2991/jegh.k.191008.001.

Moffatt, Barbara A., and Hiroshi Ashihara. 2002. "Purine and Pyrimidine Nucleotide Synthesis and Metabolism." *The Arabidopsis Book* 1. United States: e0018. doi: https://doi.org/10.1199/tab.0018.

Mule, Siva Nagi Reddy, Sharmila Nurbhasha, J. N. Kolla, Surender Singh Jadav, Venkatesan Jayaprakash, Lourdu Rani Bhavanam, and Hari Babu Bollikolla. 2016.

"Synthesis, Biological Screening and Molecular Docking Studies of Novel 4,6-Pyrimidine Derivatives as EGFR-TK Inhibitors." *Medicinal Chemistry Research* 25 (11): 2534–46. doi: https://doi.org/10.1007/s00044-016-1668-x.

Naresh Kumar, Royya, Yedla Poornachandra, Punna Nagender, Gannarapu Mallareddy, Nagiri Ravi Kumar, Palreddy Ranjithreddy, Chityal Ganesh Kumar, and Banda Narsaiah. 2016. "Synthesis of Novel Trifluoromethyl Substituted Furo[2,3-b]Pyridine and Pyrido[3′,2′:4,5]Furo[3,2-d]Pyrimidine Derivatives as Potential Anticancer Agents." *European Journal of Medicinal Chemistry* 108: 68–78. doi: https://doi.org/10.1016/j.ejmech.2015.11.007.

Natarajan, Ramalakshmi, Helina N. Anthoni Samy, Amuthalakshmi Sivaperuman, and Arunkumar Subramani. 2023. "Structure-Activity Relationships of Pyrimidine Derivatives and Their Biological Activity-a Review." *Medicinal Chemistry* 19 (1). Bentham Science Publishers: 10–30.

Nemr, Mohamed T. M., and Asmaa M. AboulMagd. 2020. "New Fused Pyrimidine Derivatives with Anticancer Activity: Synthesis, Topoisomerase II Inhibition, Apoptotic Inducing Activity and Molecular Modeling Study." *Bioorganic Chemistry* 103. Elsevier: 104134.

Poletto, Julia, Michael J. V. da Silva, Andrey P. Jacomini, Danielle L. Bidóia, Hélito Volpato, Celso Vataru Nakamura, and Fernanda A Rosa. 2021. "Antiparasitic Activities of Novel Pyrimidine N-Acylhydrazone Hybrids." *Drug Development Research* 82 (2). John Wiley & Sons, Ltd: 230–40. doi: https://doi.org/10.1002/ddr.21745.

Rashad, Aymn E., Abeer E. Mahmoud, and Mamdouh M. Ali. 2011. "Synthesis and Anticancer Effects of Some Novel Pyrazolo[3,4-d]Pyrimidine Derivatives by Generating Reactive Oxygen Species in Human Breast Adenocarcinoma Cells." *European Journal of Medicinal Chemistry* 46 (4): 1019–26. doi: https://doi.org/10.1016/j.ejmech.2011.01.013.

Rashid, Haroon Ur, Marco Antonio Utrera Martines, Adriana Pereira Duarte, Juliana Jorge, Shagufta Rasool, Riaz Muhammad, Nasir Ahmad, and Muhammad Naveed Umar. 2021. "Research Developments in the Syntheses, Anti-Inflammatory Activities and Structure-Activity Relationships of Pyrimidines." *RSC Advances* 11 (11). England: 6060–98. doi: https://doi.org/10.1039/d0ra10657g.

Reddy, Onteddu Surendranatha, Ch. Venkata Suryanarayana, K. J. P. Narayana, V. Anuradha, and B. Hari Babu. 2015. "Synthesis and Cytotoxic Evaluation for Some New 2,5-Disubstituted Pyrimidine Derivatives for Anticancer Activity." *Medicinal Chemistry Research* 24 (5): 1777–88. doi: https://doi.org/10.1007/s00044-014-1276-6.

Selvam, Theivendren Panneer, Caiado Richa James, Phadte Vijaysarathy Dniandev, and Silveira Karyn Valzita. 2015. "A Mini Review of Pyrimidine and Fused Pyrimidine Marketed Drugs." *Research in Pharmacy* 2 (4).

Shabarova, Zoe A., and Alexey A. Bogdanov. 2008. *Advanced Organic Chemistry of Nucleic Acids*. John Wiley & Sons.

Sherif, M. H., and Amal M. Yossef. 2015. "Synthesis and Anticancer Evaluation of Some Fused Coumarino-[4, 3-d]-Pyrimidine Derivatives." *Research on Chemical Intermediates* 41. Springer: 383–90.

Slagman, Sjoerd, and Wolf-Dieter Fessner. 2021. "Biocatalytic Routes to Anti-Viral Agents and Their Synthetic Intermediates." *Chemical Society Reviews* 50 (3). The Royal Society of Chemistry: 1968–2009. doi: https://doi.org/10.1039/D0CS00763C.

Soto, Ana M., and Carlos Sonnenschein. 2010. "Environmental Causes of Cancer: Endocrine Disruptors as Carcinogens." *Nature Reviews Endocrinology* 6 (7). Nature Publishing Group UK London: 363–70.

Sun, Jufeng, Cecilia C. Russell, Christopher J. Scarlett, and Adam McCluskey. 2020. "Small Molecule Inhibitors in Pancreatic Cancer." *RSC Medicinal Chemistry* 11 (2). Royal Society of Chemistry: 164–83.

Sunduru, Naresh, Anu Agarwal, Sanjay Babu Katiyar, Nishi, Neena Goyal, Suman Gupta, and Prem M S Chauhan. 2006. "Synthesis of 2,4,6-Trisubstituted Pyrimidine and Triazine Heterocycles as Antileishmanial Agents." *Bioorganic & Medicinal Chemistry* 14 (23): 7706–15. doi: https://doi.org/10.1016/j.bmc.2006.08.009.

Tang, Anqun, Keyu Gao, Laili Chu, Rui Zhang, Jing Yang, and Junnian Zheng. 2017. "Aurora Kinases: Novel Therapy Targets in Cancers." *Oncotarget* 8 (14). United States: 23937–54. doi: https://doi.org/10.18632/oncotarget.14893.

Vymětalová, Ladislava, Libor Havlíček, Antonín Šturc, Zuzana Skrášková, Radek Jorda, Tomáš Pospíšil, Miroslav Strnad, and Vladimír Kryštof. 2016. "5-Substituted 3-Isopropyl-7-[4-(2-Pyridyl)Benzyl]Amino-1(2)H-Pyrazolo[4,3-d]Pyrimidines with Anti-Proliferative Activity as Potent and Selective Inhibitors of Cyclin-Dependent Kinases." *European Journal of Medicinal Chemistry* 110: 291–301. doi: https://doi.org/10.1016/j.ejmech.2016.01.011.

Xu, Yu, Shu-Yi Hao, Xiu-Juan Zhang, Wen-Bo Li, Xue-Peng Qiao, Zi-Xiao Wang, and Shi-Wu Chen. 2020. "Discovery of Novel 2,4-Disubstituted Pyrimidines as Aurora Kinase Inhibitors." *Bioorganic & Medicinal Chemistry Letters* 30 (3): 126885. doi: https://doi.org/10.1016/j.bmcl.2019.126885.

Zhang, Jianming, Priscilla L. Yang, and Nathanael S Gray. 2009. "Targeting Cancer with Small Molecule Kinase Inhibitors." *Nature Reviews Cancer* 9 (1). Nature Publishing Group UK London: 28–39.

Zhang, Lingzhi, Qiurong Ju, Jinjin Sun, Lei Huang, Shiqi Wu, Shuping Wang, Yin Li, Zhe Guan, Qihua Zhu, and Yungen Xu. 2020. "Discovery of Novel Dual Extracellular Regulated Protein Kinases (ERK) and Phosphoinositide 3-Kinase (PI3K) Inhibitors as a Promising Strategy for Cancer Therapy." *Molecules*. doi: https://doi.org/10.3390/molecules25235693.

Zhang, Yuan, Handeng Lv, Lu Luo, Yong Xu, Yaqian Pan, Yuewu Wang, Han Lin, et al., 2018. "Design, Synthesis and Pharmacological Evaluation of N4,N6-Disubstituted Pyrimidine-4,6-Diamine Derivatives as Potent EGFR Inhibitors in Non-Small Cell Lung Cancer." *European Journal of Medicinal Chemistry* 157: 1300–1325. doi: https://doi.org/10.1016/j.ejmech.2018.08.031.

Zhao, Mingxia, Hongyu Ren, Jin Chang, Diqin Zhang, Yating Yang, Yong He, Chuanmin Qi, and Huabei Zhang. 2016. "Design and Synthesis of Novel Pyrazolo[1,5-a]Pyrimidine Derivatives Bearing Nitrogen Mustard Moiety and Evaluation of Their Antitumor Activity *in Vitro* and *in Vivo*." *European Journal of Medicinal Chemistry* 119: 183–96. doi: https://doi.org/https://doi.org/10.1016/j.ejmech.2016.04.068.

Zhong, Lei, Yueshan Li, Liang Xiong, Wenjing Wang, Ming Wu, Ting Yuan, Wei Yang, et al., 2021. "Small Molecules in Targeted Cancer Therapy: Advances, Challenges, and Future Perspectives." *Signal Transduction and Targeted Therapy* 6 (1): 201. doi: https://doi.org/10.1038/s41392-021-00572-w.

# Index

## D



## E



## F



## G



## H



## I



## L



## M



## N